Major Electrical Appliances and How to Repair Them

New Illustrated Library of Home Improvement Volume 10

Major Electrical Appliances and How to Repair Them

Prentice-Hall/Reston Editorial Staff

Prentice-Hall of Canada, Ltd./ Reston Publishing Company
Scarborough, Ontario

Series contributors/H. Fred Dale, Richard Demske, George R.
Drake, Byron W. Maguire, L. Donald Meyers, Gershon
Wheeler

Design/Peter Maher & Associates
Color photographs/Peter Paterson/Photo Design

Printed and bound in Canada.

The publishers wish to thank the following companies for
providing photographs for this volume:

John Inglis Co., Ltd.
Orli Metal Products, Ltd.
Proctor Lewyt Division of SCM (Canada), Ltd.
Ronson Products of Canada, Ltd.
Van Wyck Industries (Canada), Ltd.
Westinghouse Canada

Contents

Electric Motors

The electrical appliances discussed in the preceding volume on electrical repair are primarily devices for producing heat. To be sure, the rotisserie has a motor which turns a spit so that the food is exposed to the heat evenly, and the electric heater has a motor which rotates a fan to blow heat into the room, but the prime purpose of both of these appliances is producing heat. On the other hand, in many appliances the motor is the important element, and the motion imparted by the motor is used to accomplish other household tasks. Motors are used to drive compressors in refrigerators and to produce vacuum in vacuum cleaners, to mix food in mixers and blenders, and to rotate clothes in washers and dryers. Look around your house and you will probably be surprised at the large number of electric motors it contains.

Motors are rated according to the amount of power they deliver. Since small motors are usually somewhat less than 70 per cent efficient, it is necessary to put in 100 watts to get out the equivalent of 70 watts in mechanical energy or turning force. Motor ratings are usually given in *horsepower*, and one horsepower is equal to 746 watts. Thus, a 1/4-horsepower motor *puts out* the equivalent of 1/4 of 746 or 186.5 watts. Since this is 70 per cent of the input, the motor requires about 266 watts to drive it. Thus, it will draw about 2.3 amperes from a 115-volt line. (Watts equals volts times amperes.) On some appli-

ances with motors, the nameplate may list the power or the current instead of the horsepower of the motor. A blender, for example, may show 700 watts on its nameplate. Using the 70 per cent efficiency rule, the output of the blender motor would be 490 watts, which is about 2/3 of 746. The motor in this blender is a 2/3-horsepower motor. As another example a small fan may draw one ampere. To determine the horsepower of the fan motor, you must first figure out the power fed in, which is the product of current and voltage. In this case, one ampere times 115 volts gives 115 watts. Assuming 70 per cent efficiency, the output is about 80 watts, which is slightly more than 10 per cent of 746. This motor would be called a 1/10-horsepower motor.

Motors that put out less than one horsepower are called *fractional-horsepower motors.* All the motors used in home appliances are fractional-horsepower, ranging from about 1/20 to 3/4 of a horsepower.

Motors may be designed to run on alternating current only, on direct current only, or on either. A motor which will operate on either AC or DC is called a *universal* motor. *Induction* motors run on AC only. Most home appliances use either a universal motor or one of two types of induction motor, a *split-phase* motor or a *capacitor-start* motor. A third type of induction motor, the *shaded-pole* motor, is occasionally used in very small fans.

In selecting a motor for a particular appliance, a manufacturer must consider the properties of the motors and the requirements of the appliance, as well as cost. A motor which might work well in an electric heater would not be satisfactory in a blender. On the other hand, a blender motor could work in a heater, but would be more expensive than the motor actually used there.

Criteria for selection include speed, reversibility and torque. Should the speed be constant or variable? Can the direction of rotation be reversed? Can the motor start under a heavy load? Is the speed of rotation affected by the load? A clock motor must run at constant speed. Many other appliances have constant speed motors also, although it is not as important as in a clock. Some appliances have two or more speeds. The speed may be continuously variable, as in some mixers, or it may be possible to switch from one to another constant speed as in some fans or blenders. Appliances generally do not need reversible motors, although some motors can be reversed by switching connections on the windings. Some models of garbage disposers do have a switch for reversing direction of rotation, so that both sides of the cutting teeth can be utilized.

The *torque* is the turning force of the motor. If the motor can start under heavy load, it is said to have a *high starting torque*. Once the motor is running, it requires less force to maintain rotation. If an increase in load causes the motor to slow down, it is said to have *poor regulation*. Most orange juicers, for example, will slow down if you exert too much pressure on the orange.

Another important consideration is *starting current*. When a motor is turned on, there is a surge of current until the rotor starts to rotate. This is technically called *locked-rotor current* since it is the current with the rotor stationary. When the rotor is turning, the current drawn by the motor is much less than the locked-rotor current. A high starting current may cause lights to flicker and line voltage to decrease temporarily. With appliances which are started infrequently, the occasional flicker is tolerable, but in a refrigerator or air conditioner, which turns on and off frequently,

a high starting current would be a constant source of annoyance.

1-1. Basic Induction Motor

A motor consists of two main parts: the *stator* or stationary part, and the *rotor* or rotating part. The stator is connected to a frame for mounting. The rotor should rotate freely and is usually supported in bearings mounted in the same frame.

In a basic induction motor, the stator is a coil of wire, surrounding the rotor, but *not* in contact with it. Current through the coil creates a magnetic field which then induces current in the rotor. The rotor current also creates a magnetic field which tries to line up with the magnetic field in the coil. Since alternating current is used, the magnetic field in the coil rotates, and thus the rotor should rotate also.

There is only one difficulty with this basic motor. It won't start. If you spin the rotor after switching on the current in the coil, it would continue to rotate at constant speed as long as current flowed. It would rotate in either direction, but something must be added to start the motor. The different types of starting devices lend their names to the different types of motors.

1-2. Split-Phase Motor

A split-phase motor is a practical induction motor which uses a second coil as a starting device. The stator coils are called "field coils" since they set up magnetic fields. In the split-phase motor, the main coil is called the *running* coil or running winding, and the second coil is the *starting* coil. Their respective magnetic fields are the running field and the starting field. The two coils are displaced

Squirrel cage rotor.

magnetically, and the two-phase field causes the rotor to start spinning. After the motor has started, the starting coil is disconnected by a centrifugal switch.

The rotor of an induction motor is a solid casting with no coil windings on it. It is referred to as a *squirrel cage rotor*, and is illustrated in Figure 1-1. The ends are supported in bearings in the frame.

The stator of an induction motor could be simply a coil, but it would be inefficient. In practice it consists of pole pieces made of laminated metal strips on which the field coils and starting coils are wound, as shown in Figure 1-2. The poles for the starting coil are displaced from the running poles, as shown in the figure. The current in the starting coils is out of phase with that in the running coil so that a rotating magnetic field is produced to start the motor. The pole pieces are joined by a laminated metal ring to close the magnetic field.

After the motor is running, the starting coil should be disconnected since it is no longer needed. If the starting coil were not disconnected, the motor would still run, but it would draw current in the starting coil which would increase the power put into the motor, but

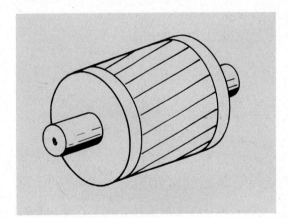

Fig. 1-1. Squirrel cage rotor.

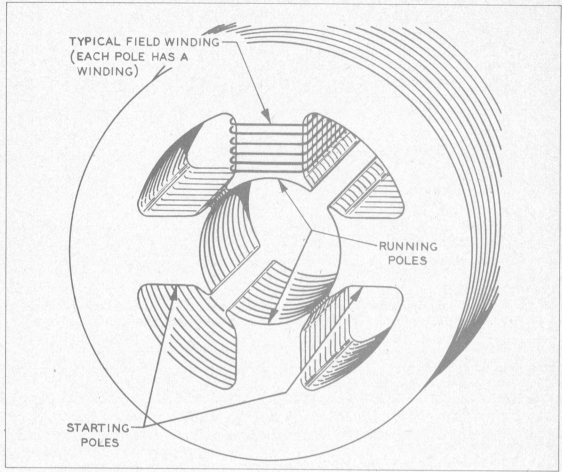

Fig. 1-2.   Pole pieces.

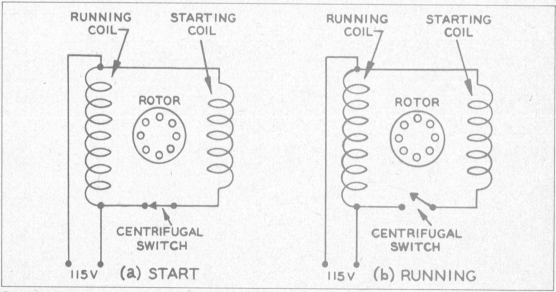

Fig. 1-3.   Split-phase motor.

would not increase the power out. Thus, the motor would be inefficient and would tend to overheat. In practice, a centrifugal switch is used to disconnect the starting coil. This switch is mounted on the rotor and is held closed by a spring. When the motor attains sufficient speed, centrifugal force overcomes the pull of the spring and forces the switch open. The switch is connected in series with the starting coil through slip-rings on the rotor.

Electrically, the circuit of a split-phase motor is as shown in Figure 1-3. In (a), both the running coil and the starting coil are connected across the 115-volt line. This is the position at rest or at the start. When the motor is running, the centrifugal switch is open, and thus the starting coil is disconnected from the 115-volt line, as in (b). The motor continues to run since the running coil is still connected. If the main switch connecting the motor to the house line is disconnected, the motor starts to

slow down and eventually stops. Even before it has reached a complete stop, the spring on the centrifugal switch will overcome the decreasing torque and close the switch, returning the circuit to that shown in Figure 1-3(a). Note that if you made a continuity check at the line input to the motor, you would have continuity even if one of the coils were open, since both are in the circuit when the motor is at rest.

In Figure 1-3, the rotor is indicated by a circle with smaller circles inside of it. This is the standard symbol for a squirrel cage rotor. Note that there are no electrical connections to the rotor in a split-phase motor. The electrical connections to the centrifugal switch, which is indeed mounted on the rotor, are not considered connections to the rotor. The rotor is usually held in place by bearings in end bells on the motor case. Frequently, long screws passing through the entire motor body join the end bells together and thus hold

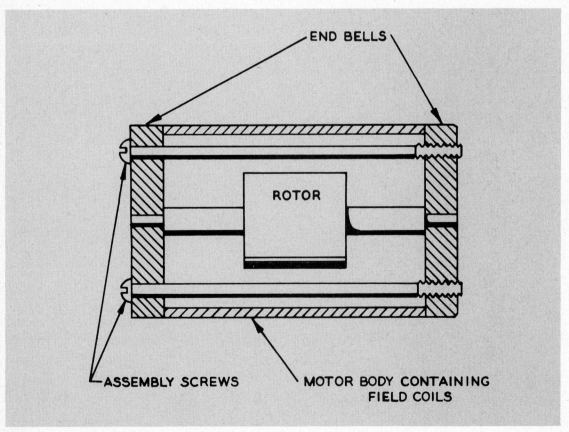

Fig. 1-4. Physical construction.

the motor together, as indicated in Figure 1-4. The rotor is not a snug fit since it must rotate freely. End play should be at least 0.005 inch, and even five or six times this amount is permissible.

A split-phase motor has a high starting current and low starting torque. However, because this type of motor is relatively inexpensive, it is used for washers, dryers and some fans. Speed of rotation is constant. The direction of rotation may be reversed by changing the coil connections when the motor is at rest.

1-3. Capacitor-Start Motor

In the split-phase motor it is necessary to have two out-of-phase fields to start the motor. This is accomplished by winding the two field coils differently. Another way of producing a different field in a second coil is by placing that coil in series with a capacitor. Schematically this is shown in Figure 1-5. Coil

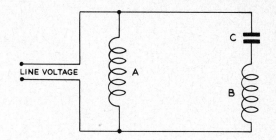

Fig. 1-5. Capacitor phase shift.

A is directly across the line. Coil B is in series with capacitor C. The capacitor introduces a phase shift across coil B, so that the two coils are out of phase.

This principal is used in a capacitor-start motor as shown in Figure 1-6. In (a), with the motor at rest, the circuit is essentially that of Figure 1-5. The different phase in the starting coil is produced by the capacitor. When the motor is running, the centrifugal switch opens, as in Figure 1-6(b). Now the capacitor and the starting coil are removed from the circuit. Note the similarities between Figures 1-5 and 1-6.

The capacitor in a capacitor-start motor is mounted outside the main motor casing. The

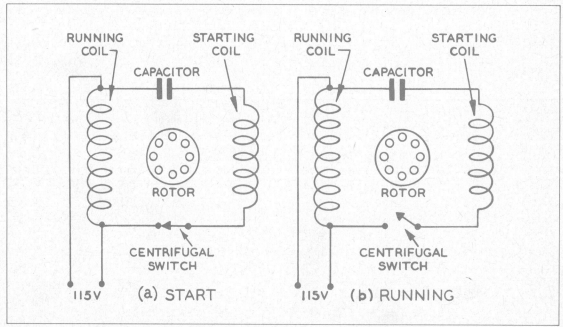

Fig. 1-6. Capacitor-start motor.

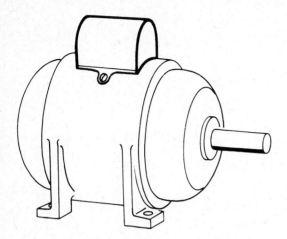

Fig. 1-7. Capacitor bulge.

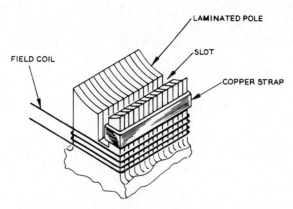

Fig. 1-8. Shaded pole.

external bulge easily identifies this type of motor. A typical shape is shown in Figure 1-7.

The capacitor on a capacitor-start motor causes an increase in starting torque (as compared to a split-phase motor) and a decrease in starting current. Thus, this motor can be used in refrigerators and air conditioners, since it does not cause objectionable light flickering. The capacitor represents an additional expense, so this motor is more expensive than the split-phase motor for the same power rating. The capacitor-start motor runs at constant speed, and direction of rotation can be changed by changing the coil connection when the motor is at rest.

1-4. Shaded-Pole Motor

Another method of producing the required phase difference is by *shading* the field pieces. This means simply slotting the field poles near one end and wrapping a copper strap around the "shaded" section, as shown in Figure 1-8. A current induced in the copper strap produces the required out-of-phase magnetic field and starts the motor. This motor differs from the others mentioned above in that the starting coil remains in the circuit after the motor is running.

The shaded-pole motor is amazingly simple and can run indefinitely. It is comparatively inefficient, usually having much less than 50 per cent efficiency. It is used in low power applications such as small fans and phonographs. Speed of rotation is constant, but direction cannot be reversed.

1-5. Universal Motor

In the induction motors discussed earlier in this chapter, the rotors are solid cast metal parts with no electrical connection to the line cord or the field coils. The universal motor works on a totally different principle and is quite different in appearance from other types of motors. Mounted on the motor frame are two field coils wound on laminated pole pieces. The armature is supported between these pole pieces (an armature is a rotor with coils wound on it). The armature is made of laminated metal strips which are slotted, and the armature coils are wound in the slots. Each armature coil is connected to a pair of opposite contacts on a commutator, but are insulated from the core of the rotor. A typical armature is shown in Figure 1-9 without coils.

The field coils in the frame and one coil in the armature are always in series. For this reason the motor is also called a *series motor.* Electrically, the circuit is as shown in Figure 1-10. Connections between the armature coil and the field coils are accomplished by brushes contacting the commutator. Thus, for any one position of the armature, one armature coil is directly connected between the

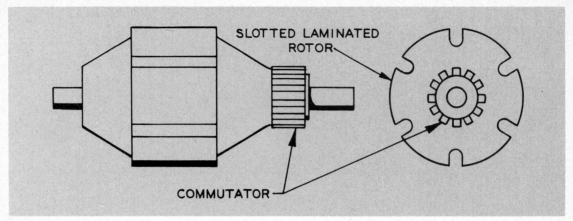

Fig. 1-9.   Armature.

opposite sections of the commutator which are under the brushes. The brushes are connected directly to the field coils. As the armature rotates, a new armature coil moves into the circuit as each one moves out, so that there is always a complete circuit through the motor.

The commutator, as indicated in Figure 1-9, is simply a slotted metal ring, with each section insulated from the next by mica strips. Each pair of opposite sections is connected to a separate coil on the armature. The coils are wound in such a manner that the magnetic field in the armature is out of line with that of the field coils. As the armature rotates to line up the two magnetic fields, a new armature coil moves into the circuit and the armature magnetic field is again out of line with that of

the field coils. As each new successive armature field is pulled into line, the armature continues its rotation.

The brushes making contact with the rotating commutator segments are carbon blocks held in contact with the commutator by a spring. A typical brush is shown in Figure 1-11. It is a rectangular block curved at one surface to make good contact with the commutator. It is held in a rectangular brush holder in which it must fit snugly but should still be free enough to be pushed by the spring to maintain contact with the commutator as the brush wears. The pigtail in Figure 1-11 is a soft, stranded copper wire which connects the brush to the field coil. If a pigtail breaks, the motor might still run since the spring is metal and can carry the current. However, the

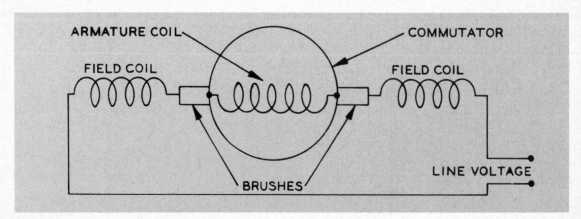

Fig. 1-10.   Circuit of universal motor.

Motor armature, commutator and brushes.

springs are not designed to carry current, and may get hot and become damaged.

The universal motor has extremely high starting torque and high speed and is used in applications where it is necessary to start a motor under heavy load. For a given horsepower, the universal motor is smaller and lighter than induction motors. The disadvantages of a universal motor are poor regulation, more noise than an induction motor, and radio interference caused by sparking at the brushes. The radio interference can be minimized by placing small capacitors across the sparking circuit, and some manufacturers furnish their appliances with these capacitors already installed.

The speed of rotation of a universal motor can be varied. There are several ways of doing this, all quite different. One method is by means of movable brushes. Recall that the brushes are positioned on the commutator to produce an armature field which is out of line with the field produced in the main field coils. There is one optimum position of the brushes

where maximum torque is produced. If the brushes are moved from this position, the motor will still run, but the reduced torque causes reduced speed. It also produces less power. The use of moving brushes is not a popular method of speed control.

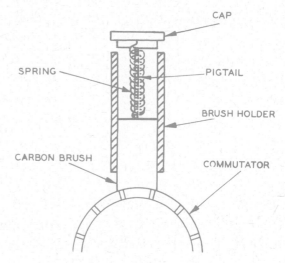

Fig. 1-11.   Brush and holder.

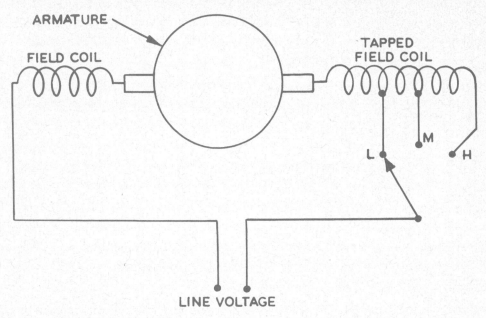

Fig. 1-12.   Tapped field speed control.

A more common method of controlling the speed of a universal motor is tapping the field coils. In effect, the magnetic field of the stator is shortened, producing less torque. This also produces less power as well as reducing the speed. A diagram of a tapped-coil circuit is shown in Figure 1-12. Only one field coil is tapped. When the switch is at *H* in the figure, the entire tapped coil is in the circuit, and the motor runs at high speed. At *M* and *L*, the coil is progressively shortened, so that the motor runs at medium speed and low speed, respectively. A rotary switch is shown in Figure 1-12, but in practice push buttons are frequently used on appliances such as blenders. It is not necessary to limit the number of taps to three.

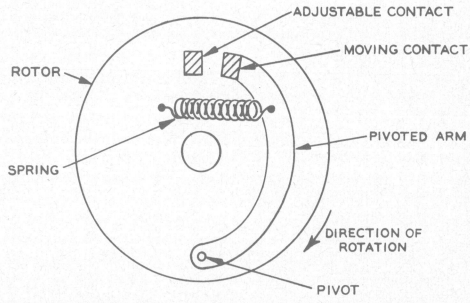

Fig. 1-13.   Centrifugal speed control.

A centrifugal speed control, shown in Figure 1-13, can be used on any kind of motor. This type of control is sometimes called a *centrifugal governor*, since when it is set it not only determines the speed of operation, but also maintains that speed under varying loads, even for motors with poor regulation. As shown in Figure 1-13, there are two contacts which are wired into the circuit so that when these two are touching, or closed, field current must pass through their connection. When they are separated, or open, the circuit is broken, and no current reaches the armature coils. For example, the switch may be in series with one of the brushes. One contact is on the end of an arm which is free to rotate around a pivot point near the other end of the arm. This is the *moving contact.* The other contact of the switch is fixed, although its position is adjustable. When the motor is at rest, a spring pulls the moving contact against the adjustable contact. Now if the motor is turned on, it starts in its normal manner. As it speeds up, it reaches a point where the centrifugal force overcomes the pull of the spring and separates the two contacts, thus breaking the circuit. With no voltage supplied, the motor now slows down. The centrifugal force is reduced, and the spring is able to pull the contacts together again. The motor speeds up, and the process is repeated. This making and breaking of the circuit occurs many times

in one revolution of the motor, so that the motor runs smoothly. The speed is set by moving the adjustable contact. When it is moved to the left in Figure 1-13, centrifugal force will break the contact at a lower speed than when the contact is moved to the right. In effect, the voltage is off a greater portion of the time to produce a lower speed.

In order to prevent pitting of the contacts from the frequent making and breaking of the circuit, it is necessary to include a small capacitor across the contact points. This also minimizes radio interference. The centrifugal switch is more costly than a tapped-coil speed control because of the added parts.

The direction of rotation of a series motor can be reversed by reversing the armature connections. One way of doing this is shown in Figure 1-14. Two switches are shown but in practice a single double-pole double-throw switch is usually used. Assume that at any one instant current enters the field coil at the left and leaves from the field coil at the right, as it does half the time when AC is used. With the switches in the positions indicated by the solid arrows, current flows through the armature coil from left to right. Assume that the magnetic field set up in the armature is then pulled toward the pole pieces by their magnetic fields. Now if the switches are moved to the positions indicated by the dotted arrows, the current through the armature coil flows from right to left. This produces a magnetic

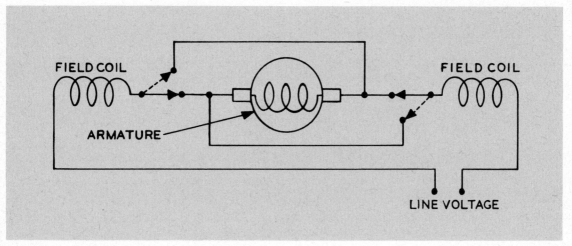

Fig. 1-14.   Reversing switch.

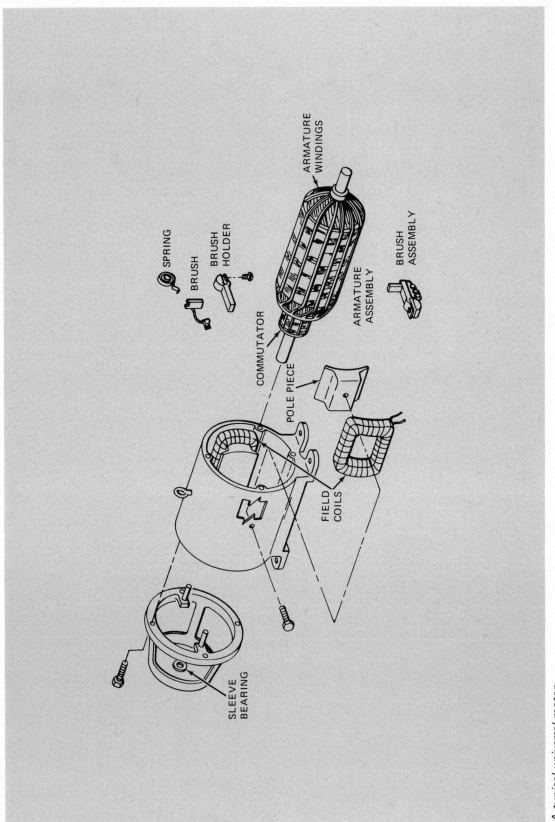

A typical universal motor.

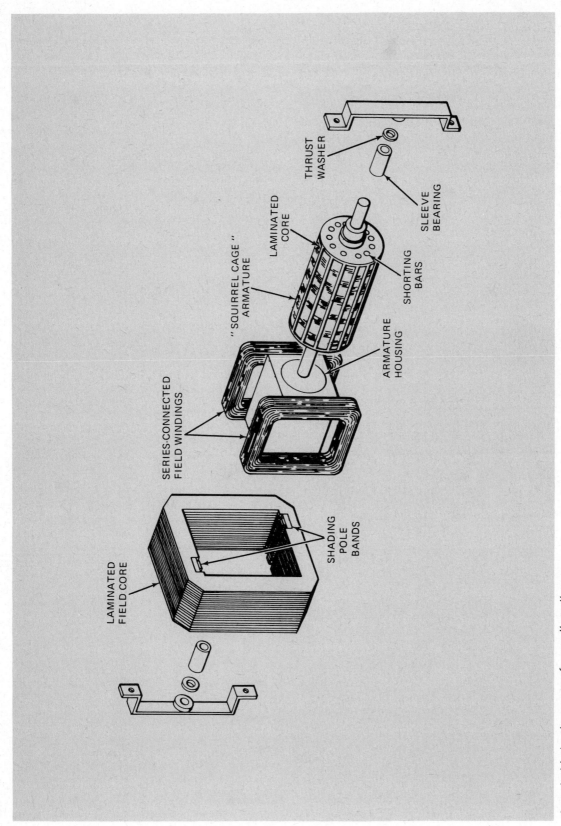

THRUST WASHER

SLEEVE BEARING

LAMINATED CORE

"SQUIRREL CAGE" ARMATURE

SHORTING BARS

SERIES-CONNECTED FIELD WINDINGS

ARMATURE HOUSING

LAMINATED FIELD CORE

SHADING POLE BANDS

A typical induction motor for small appliances.

field in the armature which is opposite to the first field, and thus, the armature would be repelled where before it was attracted. The armature thus turns in the opposite direction.

Because of its very high starting torque under heavy load, a series motor is used to run a streetcar. In the home it is used for vacuum cleaners, mixers, blenders, sewing machines, power tools and in other appliances which start and run under *varying* loads. The universal motor is capable of very high speed, and the speed can be controlled by any of the methods described earlier.

1-6. Summary of Motor Characteristics

The shaded-pole motor is exceedingly simple and rarely requires servicing. It has very low starting torque and runs at constant speed. It may be used for very low-power applications where the load is minimal.

The split-phase induction motor has low starting torque and high starting current. It runs at constant speed. It is used in medium to low power applications in appliances which are not turned on and off frequently. The high starting current causes flickering of lights whenever a split-phase motor is turned on.

The capacitor-start motor has medium torque, medium starting current, and runs at constant speed. It will supply more power than the split-phase motor. It may be used in appliances which are turned off and on frequently, since its medium starting current lessens the annoyance of light flicker. Because of the capacitor, this type of motor is more expensive than a comparable split-phase motor.

The universal or series motor has very high starting torque and can start under heavy load. It has very low starting current. The speed varies and depends on the load. It is used in appliances which may start and run under varying load conditions. For a given size and weight, the universal motor furnishes more power than any of the other motors described above.

Some typical applications of motors are presented in Table 1-1.

Table 1-1. Motor Applications.

Appliance	Typical rating (horsepower)	Motor type
small fan	less than 1/20	shaded-pole
medium fan	up to 1/3	split-phase
large fan	1/4 1	capacitor-start
mixer	1/10	universal
blender	2/3	universal
vacuum cleaner	1/4 3/4	universal
juicer	1/6	universal
washer	1/4 1/2	split-phase
refrigerator	1/6	capacitor-start
phonograph	1/20	shaded-pole
hair dryer	1/20	shaded-pole
sewing machine	1/4 3/4	universal

1-7. Repairs

Although there are many different kinds of motors, they suffer similar ailments, and the techniques for locating a source of trouble are much the same for all of them. There are some differences, of course. For example, induction motors do not have brush troubles, and universal motors are not troubled by a shorted starting capacitor. However, most apparent motor troubles are usually not in the motor itself, but in the equipment connected to the motor. Although motors are simple to take apart by removing the assembly screws shown in Figure 1-4, you should not disassemble a motor until you have eliminated all sources of trouble that can be corrected externally.

When an appliance containing a motor fails to operate, first make sure that the outlet has voltage and that the line cord and main switch of the appliance are in working order. The most frequent trouble in vacuum cleaners, for example, is a defect in a line cord caused by constant flexing or tugging. Use the techniques described in Chapter 3, Volume 4, to check the line cord and repair it if necessary. A simple continuity test can be used to check the main switch. If it is defective, replace it with a similar part.

Assuming the line cord and switch are in working order and voltage is getting to the motor, troubleshooting depends on the symptoms first and only secondarily on the type of motor. There are many different kinds of symptoms:

1. Nothing happens when a motor is turned on. It won't start and is silent.

2. The motor won't start, but it hums or tries to start.

3. The motor runs, but gets hot.

4. The motor is noisy.

5. In a universal motor, the brushes spark.

6. The motor runs, but has no power.

1-8. Nothing Happens (1)

If a motor won't start and is silent, even though voltage is reaching it, the trouble is probably in the motor. However, many motors have a *thermal overload protector*, which is a switch that shuts the motor off when it gets too hot. Before the motor can be started again, this switch must be closed manually, which is done by pushing a *reset* button on the outside of the motor case. Look for such a button (it's usually red) and push it before making any other test. If there is no thermal overload protector, or if the motor still won't start after you push the reset button, it is necessary to consider the type of motor. In any motor, a burnt-out winding will prevent starting.

In a split-phase motor, the trouble could be an open field coil, an open starting coil, or a centrifugal switch stuck in the open position. Look again at Figure 1-3, repeated here for convenience. Note that a continuity check across the line (with the power disconnected, of course) would not tell you anything, since there would be continuity even if one of these three elements were open. However, if the running coil is good, the motor should run once it is started. Plug in the motor and spin the rotor with your fingers. If the running coil is good, the motor will run, and therefore the trouble is in the centrifugal switch or the starting coil. If the motor still does not run, the trouble is probably in the running coil (field coil). With the power disconnected, hold the centrifugal switch open with your fingers or a piece of cardboard and make a continuity check across the field coil. You will probably

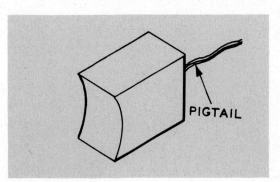

Fig. 1-15. Brush.

find it open. It is possible to buy a new field coil assembly to replace a defective one, but if the motor is old, it pays to replace the whole motor if a coil is open. In rare cases, you may be able to spot the break and solder the ends together, but it is a difficult job for an amateur, and a motor repairman would charge more than a new motor costs.

If the motor runs after you spin it by hand, check the centrifugal switch. With the power disconnected, you should be able to open and close this switch easily with your fingers. If the contacts are dirty or pitted, they may not be making an electrical connection. They can be cleaned with sandpaper. Note: *never use emery paper*, since the grit is conducting and can cause a short circuit if any particles fall into the motor. A quick check on a centrifugal switch can be made by shorting it with a clip lead. If the trouble is caused by an open switch, the motor will start with the clip lead removing the switch from the circuit. If the centrifugal switch is stuck in the open position, clean the moving parts and add a drop of oil on its pivot point. Be careful not to use too much oil, since excess oil may leak between the contacts of the switch and form an insulating coating which prevents the switch from closing.

If the centrifugal switch is good, but the motor will not run unless you spin it with your fingers, the trouble is probably an open starting coil. You can make a continuity check across it, holding the centrifugal switch open with your fingers or a piece of cardboard. If the starting coil is defective, it doesn't pay to fix it. It can be replaced, or if the motor is old, it is better to replace the whole motor. In general, it doesn't pay to try to fix burnt-out or open coils.

In a capacitor-start motor that won't start, the trouble could be any of those areas described above for the split-phase motor, and in addition it could be caused by a defective capacitor. Try spinning the motor by hand to start it. If it now runs, the trouble could be in the capacitor, the centrifugal switch, or either of the windings. Look again at Figure 1-6. The capacitor should look like an open circuit or a very high resistance. Check the two coils for continuity (as described above

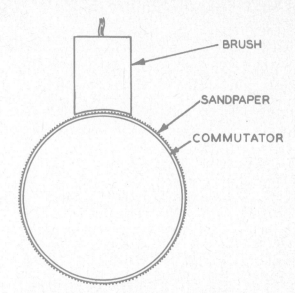

Fig. 1-16.   Fitting brush to commutator.

for the split-phase motor) and make sure the centrifugal switch is operating properly. As before, you should be able to move it easily with your fingers, and the contacts should be clean. If the capacitor is open, it cannot be checked by a continuity test since it would show no continuity when it is good as well as when it is open. Disconnect the capacitor and put it across a voltage source for a few seconds. If the capacitor is good, when you remove it from the voltage source, there will be a voltage stored in the capacitor. Now bring the leads of the capacitor together, and they should produce a spark when they touch. Use insulated pliers to hold the capacitor leads when you do this. If there is no spark, the capacitor is probably defective and should be replaced. Since capacitors are a frequent source of trouble, mail order houses carry replacement capacitors for many appliances.

If a universal motor won't start, the trouble is usually in the brushes. By referring to Figure 1-10, you can see that a continuity test can be used to check a universal motor. Occasionally the trouble will be a worn-out brush. Check to make sure brushes are being held in contact with the armature. If they are worn, replace them. Note that one end of the brush is curved to approximately the shape of the commutator, as shown in Figure 1-15. When new brushes are installed, they will spark for a few minutes until they are worn to an exact fit on the commutator. Don't worry about this sparking for the first five minutes, but if it persists much longer, make sure the brushes are not too loose in their holders. If the curved part of the brush does not mate with the commutator, you can grind it in, as shown in Figure 1-16. Wrap a piece of sandpaper around the commutator with the grit side out. Then put the brushes in the holders and turn the commutator by hand. The sandpaper will wear away the brush until there is a good fit. The brushes in motors in most appliances can be removed from the outside without taking the motor apart.

If a field coil is open, you should replace the whole stator. These open coils can be located by the usual continuity checks. In the armature, make continuity checks across each pair of opposite segments of the commutator. Each pair is attached to a different coil, and if one of them is open, the whole armature should be replaced. Also check continuity between each commutator segment and the body of the armature. You should read an open circuit here, since continuity would indicate a coil shorting to the body. If an armature has a shorted coil, replace the whole armature. Don't try to rewind the coil. It's not worth the effort, since a new armature is relatively inexpensive.

1-9. The Motor Hums but Won't Start (2)

In any kind of motor, tight bearings, a bound load, or other mechanical trouble can prevent the motor from starting, although it will hum and try to start when it is turned on. Note: *disconnect it quickly,* since a motor can burn out under these conditions. A simple check is to try to turn the motor by hand *with the power off.* It should turn easily. If the rotor cannot be moved easily, there are three possible sources of trouble. First, check the load. To do this, remove any belts or gears connecting the motor to the load. In a blender, remove the glass bowl. Try to turn the motor again. If it now rotates easily with the load detached, the fault is not in the motor, but in the load. Look for mechanical troubles, such as bent linkages, tight or broken bearings, rust or dirt. Clean and repair the load as required.

If the motor is still bound with the load detached, the trouble is caused either by the rotor rubbing against the stator or by a tight or frozen bearing. When a bearing is tight or frozen, it may simply need cleaning or oiling. Be careful not to use too much oil since an excess can spatter, and if oil drops get on the contacts of the centrifugal switch or between brush and commutator, they can break the electrical circuit. If a bearing cannot be cleaned or remains bound, replace it. When cleaning or replacing a bearing make sure you observe exactly how the washers and shims are positioned on the end of the spindle, and when you reassemble the motor, put them back the same way. These shims prevent too much end play in the armature so that the brushes are positioned correctly on the commutator.

If the rotor rubs against the stator, a broken bearing may be the culprit. Also, an end bell may be bent or broken. In either case replace the defective part with one that is identical. Motor parts are available from authorized service dealers or from the manufacturer at the address on the nameplate.

If an induction motor has a free-spinning rotor but still hums without starting, the trouble can be in the centrifugal switch. The switch may not be closing perfectly because of dirty contacts. Clean the switch contacts with sandpaper. Make sure the parts of the switch can be moved freely. If the rotor has too much end play, the vibration can prevent the switch from acting properly. End play should be at least 0.005 inch and may be up to five or six times this amount. Note that an open centrifugal switch was also given as a cause of failure if the motor didn't start and was silent. The difference here is that the points are in close enough contact to move the rotor, but the current separates the points before the rotor moves. In this case, as before, you can start the motor by spinning it, and then it will run.

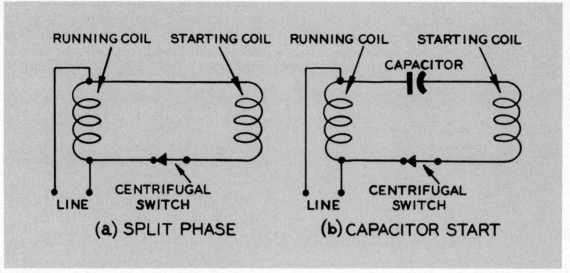

Fig. 1-17.   Comparison of induction motors.

In a capacitor-start motor, the motor can still run if the capacitor is shorted. This is shown in Figure 1-17. In (a), you see the electrical circuit of a split-phase motor, and in (b) that of a capacitor-start motor. Note that if the capacitor is shorted in (b), the circuit becomes identical with that of the split-phase circuit in (a). Recall, however, that the capacitor was added to give higher starting torque, and the capacitor-start motor is usually used in applications where this extra starting torque is needed. Thus, when the capacitor is shorted, the motor no longer has high starting torque. Since this type of motor is usually connected to a heavy load, the shorted capacitor prevents it from turning the load, although it strains to do so, and hums while trying. The solution obviously is to replace the capacitor.

1-10. The Motor Overheats (3)

Tight bearings or a heavy load can cause a motor to overheat. When the motor has cooled, disconnect it and try spinning the rotor with your fingers. If it fails to turn easily,

look for bearing trouble, rotor rubbing against stator, or load trouble. These faults and their remedies are discussed in the two preceding sections.

If the rotor spins easily, but the motor gets hot, a possible trouble is a centrifugal switch which remains closed. Referring to Figure 1-3 you can see that when the motor is in the start position, two coils are across the line. After the motor is running, the centrifugal switch is supposed to open, as in Figure 1-3(b), and only one coil remains. If the switch fails to open, the starting coil remains in the circuit and continues to draw current. This excess current causes overheating. To remedy this, make sure the centrifugal switch can be moved easily with your fingers. If the contacts are burnt shut, separate them and clean them with sandpaper. *Never use emery paper*, since the dust from emery paper conducts electricity and can cause a short circuit if it falls in the motor. Make sure the parts of the switch move easily. If necessary, oil the pivot lightly. Warning: too much oil is bad since the excess can leak between the contacts and prevent them from closing.

If a motor gets too dirty, it will tend to overheat. Many motors have a built-in fan for cooling, but if the air intake slots are clogged, the fan will labor in vain. Blow out or vacuum the motor and make sure all air vents are open. An old toothbrush can be used to clean accumulated dirt off the stator. Never use any sort of cleaning liquid or water on the coils or internal parts of the motor.

Anything that causes a motor to draw excessive current is bad, since excessive current causes overheating. If stator windings become shorted or grounded to the case, the motor will run but will draw excessive current and overheat. If you have an ammeter, you can measure the current and compare it to the rating on the nameplate. Without an ammeter you can still get a good idea of the current being drawn by observing the electric meter which registers the amount of electricity used. This meter is located outside your house. A typical meter is shown in Figure 1-18. In the center of the face there is a slot, and a horizontally mounted rotating disc protrudes through this slot. Whenever electric current is being drawn anywhere in the house, this disc rotates, and the speed of rotation is directly proportional to the current. Make sure that all electricity in the house is off. Disconnect all appliances such as refrigerators or furnaces which may start of their own accord. You will then note that the disc on the meter is stationary. Now turn on something with a known current or power rating. For example, turn on a 200-watt lamp. You will see that the disc is now rotating. There is a black smudge at one point of the disc's circumference, and by counting the number of times this smudge passes you can count the revolutions of the disc. Assume the disc makes two revolutions per minute when the 200-watt lamp is on. Now if you turn off the lamp and turn on a 1000-watt toaster, you would be using five times as much current and the disc would turn five times as fast or ten revolutions per minute. Since you know that power in watts is volts times amperes, you can calculate that a 200-watt lamp will draw slightly less than two amperes from a 115-volt line. Suppose a motor is rated to draw six amperes. You would expect it to make the disc rotate about three times as fast as the lamp or about six revolutions per minute. Actually, if it rotates seven or eight times per minute it would not be excessive, but ten or twelve revolutions per minute would indicate that the motor is drawing excessive current. If you find that the motor is drawing too much current, check the coils for shorts or grounds. If coils are bad, they should be replaced.

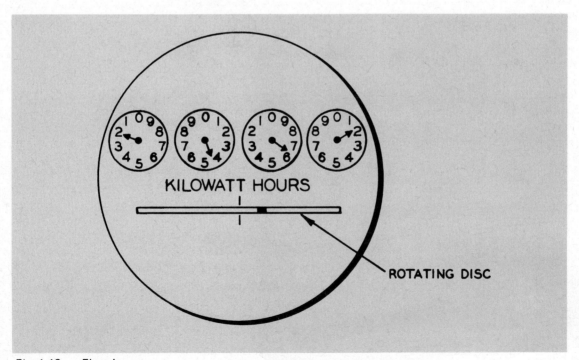

Fig. 1-18.   Electric meter.

1-11. The Motor Is Noisy (4)

Noise in a motor is usually a mechanical problem. Most likely something is loose. Go over the motor carefully and tighten all housings, baffles and other attachments. Noise can also be caused by excessive end play of the rotor or by a loose bearing. Tighten bearings if needed. Possibly a shim or washer was omitted in assembly. Check the end play and add small thrust washers to reduce the end play to something between 0.005 and 0.025 inch. Noise may also be caused by loose parts in the load, such as a loose belt, pulley, gear or fan. This is simple to track down, and corrective measures are obvious.

Dirt inside a motor can also cause noise, especially if the dirt gets between the stator and rotor. Dirt includes floor dirt as well as dust and lint. In mixers especially, all kinds of unwanted materials accumulate in the motor.

To clean a motor, first take it apart and blow off all the loose dirt or use a vacuum cleaner to pick it up. Make sure the vacuum cleaner doesn't pick up loose screws, washers, or other parts of the motor. Use an old toothbrush or a wooden tool to clean dirt from the windings. Never use liquid on any part of the motor if the liquid can get to the windings. Bearings, shafts and the centrifugal switch can be cleaned in solvent if they are first detached from the motor. Make sure they are dry before reassembling the motor. Put a drop of oil on each bearing before reassembling.

1-12. The Motor Is Sparking at Brushes (5)

Excessive sparking at the brushes of a universal motor is a warning signal of trouble. If there is not something already wrong with the motor, there soon will be. Fortunately, the

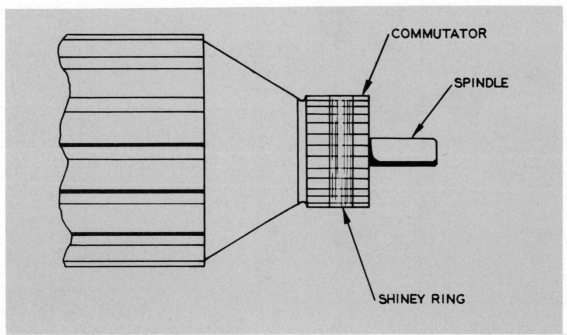

Fig. 1-19.   Armature with good commutator.

trouble is usually in the brushes themselves and is easily remedied. Remove the brushes and check the way they fit against the commutator. If the fit is bad, put sandpaper around the commutator, as shown in Figure 1-16 and discussed above, and rotate the armature until the brushes are sanded to shape. If a brush does not slide easily in the brush holder, there will also be sparking, due to poor contact between the brush and the commutator. Also, if the brush is worn down so that the spring no longer pushes it against the commutator, or if the spring itself has lost its tension, there will be sparking. Put in new brushes or springs or both, as required. New brushes must fit snugly in the brush holders, but should slide easily. If they bind, you can sand down the rough spots.

A more serious cause of sparking at the brushes is a shorted or open coil on the armature. If the brushes seem to be in good condition, check continuity between opposite segments of the commutator and between each segment and the body of the rotor. If the armature has a defective coil, replace the armature. *Do not* try to fix the coil.

The commutator itself can be damaged by poor brushes and can then cause trouble after the bad brushes are replaced. When a motor runs properly, there will be a shiny ring around the commutator where the brushes make contact. This is shown in Figure 1-19. If one segment of the commutator protrudes from the surface, it will push up the brush which then bounces on the next segment and causes sparks. These sparks burn pits in the commutator and later, when the high segment is fixed, the arc burns can cause more arcing. Such a pitted segment is shown in Figure 1-20. Similarly, if there is arcing caused by a bad brush, there will be pitting all around the commutator. After the brush is replaced, there may be arcing caused by the pits. Whenever new brushes are added, the commutator should be cleaned and sanded if necessary.

When a commutator has seen many years of service the shiny ring may become a groove around the commutator, as shown in

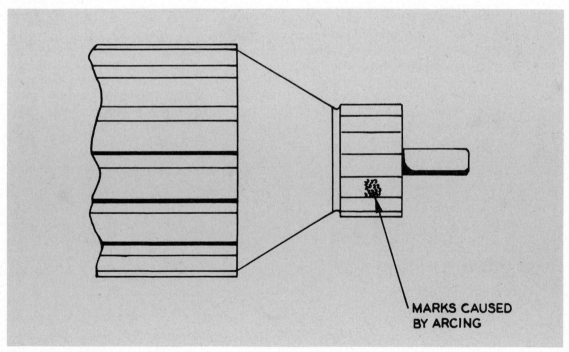

MARKS CAUSED
BY ARCING

Fig. 1-20.   Pitted commutator.

Reference Guide to Probable Causes of Motor Troubles

MOTOR TYPE	A. C. SINGLE PHASE				A.C. POLYPHASE (2 or 3 phase)	BRUSH TYPE (Universal, Series, Shunt or Compound)
	SPLIT SPLIT PHASE	CAPACITOR START	PERMANENT-SPLIT CAPACITOR	SHADED POLE		
TROUBLE	*PROBABLE CAUSES					
Will not start	1, 2, 3, 5	1, 2, 3, 4, 5	1, 2, 4, 7, 17	1, 2, 7, 16, 17	1, 2, 9	1, 2, 12, 13
Will not always start, even with no load, but will run in either direction when started manually	3, 5	3, 4, 5	4, 9		9	
Starts, but heats rapidly.	6, 8	6, 8	4, 8	8	8	8
Starts but runs too hot.	8	8	4, 8	8	8	8
Will not start, but will run in either direction when started manually—over heats	3, 5, 8	3, 4, 5, 8	4, 8, 9		8, 9	
Sluggish—sparks severely at the brushes.						10, 11, 12, 13, 14
Abnormally high speed—sparks severely at the brushes.						15
Reduction in power—motor gets too hot.	8, 16, 17	8, 16, 17	8, 16, 17	8, 16, 17	8, 16, 17	13, 16, 17
Motor blows fuse, or will not stop when switch is turned to off position.	8, 18	8, 18	8, 18	8, 18	8, 18	18, 19
Jerky operation—severe vibration						10, 11, 12, 13, 19

*PROBABLE CAUSES
1. Open in connection to line.
2. Open circuit in motor winding.
3. Contacts of centrifugal switch not closed.
4. Defective capacitor.
5. Starting winding open.
6. Centrifugal starting switch not opening.
7. Motor over-loaded.
8. Winding short circuited or grounded.
9. One or more windings open.
10. High mica between commutator bars.
11. Dirty commutator or commutator is out of round.
12. Worn brushes and/or annealed brush springs.
13. Open circuit or short circuit in the armature winding.
14. Oil-soaked brushes.
15. Open circuit in the shunt winding.
16. Sticky or tight bearings.
17. Interference between stationary and rotating members.
18. Grounded near switch end of winding.
19. Shorted or grounded armature winding.

(Courtesy *Bodine Electric Company*)

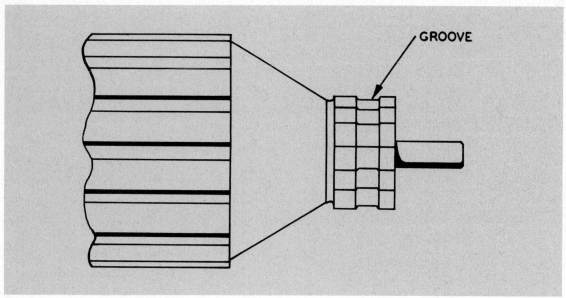

GROOVE

Fig. 1-21.   Grooved commutator.

Figure 1-21. This is not necessarily bad. However, if the surface falls below the level of the mica insulators, the brushes may bounce on the mica. Use a hacksaw blade to cut off any mica that protrudes above the surface of the commutator where the brushes might strike. Also, whenever a commutator is sanded to remove pit marks, make sure the mica insulators do not project above the surface. Use a hacksaw blade to trim off excess mica.

1-13. The Motor Turns but Has No Power (6)

If a motor runs, but lacks power, first look for mechanical troubles. With power off, make sure the motor and its load are free to move without restriction. Bearings which are too tight or need oil can hamper operation. Also, if the end play is less than the minimum 0.005 inch, the expansion due to heat when the motor is running can cause the rotor to bind. If the motor does not turn easily, look for a mechanical defect in bearings, belt, gears, fan and other likely places, and loosen or replace parts until the motor turns easily.

If the capacitor is shorted in a capacitor-start motor, the motor will lack power on starting, but once it is running, the centrifugal switch will remove the capacitor from the circuit, and the motor should run satisfactorily. On any induction motor, if the centrifugal switch fails to open, the motor will be hot and inefficient and will draw too much current.

As a final check in any motor, make sure there is no voltage on the frame of the motor while the motor is running. You can do this by putting one probe of your meter or test-light on the frame and the other on ground. Voltage on the case is a shock hazard, but in addition it may be a sign of a coil shorted to the case, which would eventually cause failure.

Control Devices

Automatic appliances such as dishwashers and clothes washers perform a number of different functions, such as washing, rinsing and drying. These functions must be performed in the correct order. Thus, in addition to the switches which start and stop each of the functions, there must be some sort of "brain" which determines when each event shall occur. The switches are control devices in that they control individual functions. The "brain" is also a control device since it controls the switches.

In a sense every electrical appliance needs at least one control device, even those with no switches. For example, if a popcorn popper starts to heat up when you plug the cord in an outlet, the control is plugging it in. Most appliances, however, do have switches to start them, and these switches take a variety of shapes and forms. In appliances which are connected to plumbing there are also mechanical switches to start and stop the flow of water, but these mechanical switches are usually controlled electrically.

In an automatic appliance, the master control determines which switches turn on and off. The determination may be based on time or on a physical condition, such as water reaching a suitable level or an oven reaching a desired temperature. Temperature sensing switches, or thermostats, are considered in detail in Chapter 4, Volume 4. They are used in most heating appliances. Other control devices, switches, timers and solenoids, are discussed in this chapter.

2-1. Switches

Consider the many different ways you "turn on" electricity. The common wall switch for lighting is flicked up or down. Some blenders have a set of push buttons for selecting the proper speed. Push buttons are also used on some heaters to select the proper temperature. The tip-over switch on a heater is a push button type of switch which must be held in to keep the electricity connected. Another type of push button switch is found on many electric juicers. It is pushed and released to turn the appliance on, and again pushed and released to turn it off. A tank-type vacuum cleaner usually has a switch which is operated by stepping on it. This is essentially a foot-operated push button switch. Some appliances have a rotary switch, as, for example, a dishwasher or clothes washer. Some mixers have a sliding lever, where the amount of movement regulates the speed of the mixer. A toaster is turned on by depressing the lever which lowers the bread.

Regardless of their external form, all switches have one thing in common: they simply connect two contacts when they are in

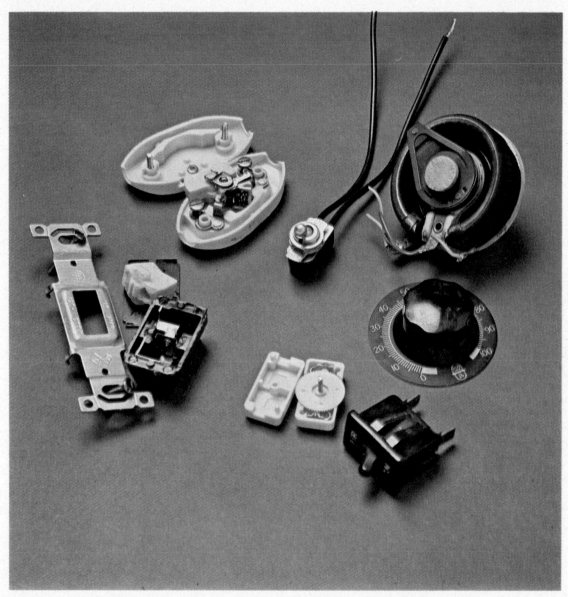

*All switches have one thing in common: they connect two contacts when in the **on** position.*

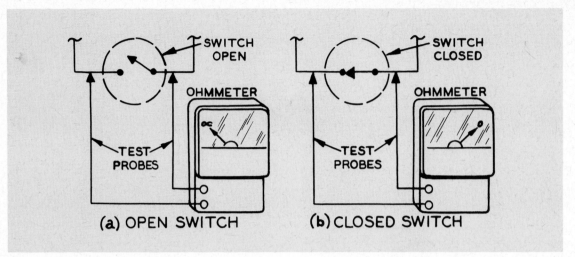

Fig. 2-1. Continuity across switch.

the *on* position and separate the contacts when they are *off*. If you checked continuity across a switch, it would look like an open circuit when it is off and like a short circuit when it is on, as shown in Figure 2-1. In (a), the switch is open, and thus when the probes from the ohmmeter are placed across it, the meter indicates infinite resistance , which means an open circuit. In (b), with the switch closed, the meter indicates zero ohms, or a short circuit. When checking the continuity of a switch, you must make sure that there is no other continuous path across the switch in the rest of the circuit.

A switch can fail by sticking in the open or closed position. If this happens, the switch should be replaced. Disconnect all wires to the switch, noting carefully where each wire is connected, and put in a new switch. Connect the wires to the new switch in the same manner as they were connected to the old switch. Sometimes one switch in an assembly of several push button switches goes bad, and it may be necessary to replace the whole assembly because the switches cannot be separated. Here, many wires have to be disconnected and reconnected, and you should make a drawing of the connections before disconnecting a single wire.

A switch that simply makes and breaks one connection, like that in Figure 2-1, is called a *single-pole single-throw switch*. "Single-pole" refers to the fact that only one wire is

connected or disconnected by the switch. "Single-throw" means that there is only one method of connecting this wire. In more complicated switches, one or more wires may be connected at different times to one or more contacts. For example, in a rotary switch, the wire connected to the rotating contact may be connected to any of several contacts in turn as it is rotated. A *single-pole double-throw* switch, which might be used to select a high or low heat in a heating appliance, is shown in Figure 2-2. Note that with the switch in the center position shown, the appliance is off. If the switch is moved to the *H* contact, only the high heating element is in the circuit. In the *L* position, only the low element is connected. There are two *on* positions, hence "double-

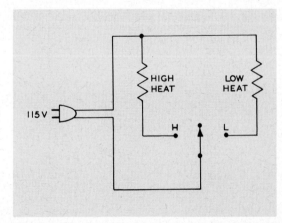

Fig. 2-2. Single-pole, double-throw switch.

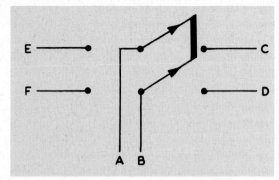

Fig. 2-3. Double-pole, double-throw switch.

throw''. Externally the single-throw and double-throw switches may appear identical, but the double-throw has three contacts whereas the single-throw has two.

When checking a double-throw switch for continuity, you must take into careful consideration what you are doing and what the rest of the circuit is. For example, in Figure 2-2, if you measured the continuity between the *H* and *L* contacts, you would always read the sum of the two heating resistances, regardless of the position of the switch. To check this switch it should be disconnected from the rest of the circuit first. Now there should be an open circuit between two of the contacts regardless of the position of the switch. These would be the ''throw'' contacts, shown as *H* and *L* in the figure. Between the third or *common* contact and either of the others there should be an open circuit in one position

of the switch and a short in the other. If the switch is bad and needs replacing, make sure that the common wire is connected to the common contact of the new switch.

A *double-pole, double-throw* switch is shown schematically in Figure 2-3. When the switch is closed to the right, wire A is connected to wire C, and wire B to wire D. When the switch is moved to the left, A is connected to E, and B to F. In effect, there are two switches which are tied together physically so that both are thrown simultaneously by moving only one lever. An application of the double-pole, double-throw switch is shown in Figure 1-14, as a reversing switch for a series motor. In this application the switch is thrown to one pair of connections or the other. In other applications, there may be a *neutral* or *off* position between the two, but the switch is still double-throw.

A common application of a single-pole, double-throw switch is to control a ceiling light fixture from two switches at either end of a room. Two such switches are used, as shown in Figure 2-4. When both switches are connected to the same wire running between them, the circuit is closed and the lights light. If either switch is thrown to the other wire, there is no continuous circuit and the lights go out. Note that from the off position, the lights will go on if either switch is moved.

An unusual application of switching is in an electric range. Using two identical Calrod

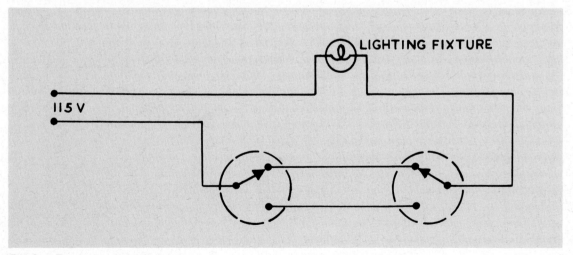

Fig. 2-4. Two controls for lighting.

Timers are master controls that open and close switches at the proper time.

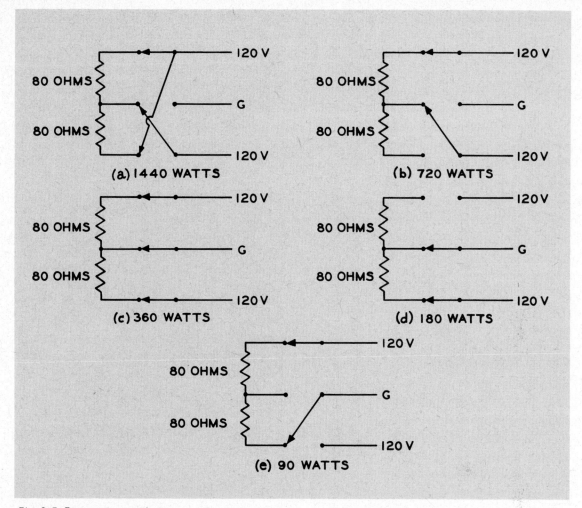

Fig. 2-5. Range connections.

elements in one heating unit, with appropriate switches, it is possible to get five different degrees of heat. This is indicated in Figure 2-5 where two 80-ohm elements are shown. The range is serviced by a three-wire line, where one line is at ground and each of the other two are 120 volts from ground. The phases of the two hot lines are such that they differ by 240 volts.

In (a), both 80-ohm elements are in parallel across 240 volts. Each draws three amperes, and the total power in both is 1440 watts. In (b), the 240-volt line is across one 80-ohm element to produce 720 watts. In (c), each of the elements is across 120 volts and draws 1.5 amperes. The total current of three amperes produces 360 watts. In (d), only one element is used across 120 volts, producing 180 watts. Finally in (e), both elements are in series across the 120-volt line. The 160-ohm resistance draws 0.75 ampere, and the power is only 90 watts.

2-2. Three-way Lamp

A very common appliance found in most homes is a three-way lamp. In one version, two or more bulbs are switched on separately or in combination. In another, a single three-way bulb with two filaments is used. Either of these filaments can be switched on separately or both together. Two popular types of three-way bulbs are the 100-200-300 watt and the 50-200-250 watt. In both the multiple bulb version and the three-way bulb, the degree of lighting is controlled by a switch. This is usually a rotary switch with a cam-like blade, as shown in Figure 2-6. The three-way bulb itself has two filaments. In the 100-200-300 watt bulb, the low filament draws 100 watts and the high filament draws 200 watts. In the 50-200-250 watt bulb, the low and high filaments draw 50 and 200 watts, respectively. In the figure the two contacts are numbered 1 and 2. With the switch blade as shown in Figure 2-6, neither filament is connected, and the lamp is off. As the switch is rotated, detents stop it at the proper contact positions. In the first position, the narrow part of the blade contacts point 1 and closes the circuit to the low filament. In the second position, the narrow blade touches point 2, and the opening between the blades falls in the vicinity of point 1. Thus, only the high filament is connected. In the third position, the wide blade touches both points 1 and 2, and both filaments light. Finally, the next position is a return to the off position

shown. In the lamp with separate bulbs, the switch operates in the same manner.

A three-way lamp can develop trouble in the switch itself or in the contacts to the bulb. If the switch is noisy, it indicates arcing inside, usually a sign that the switch will soon fail. As with most switches, it doesn't pay to try to repair it. Simply replace the switch with a new one that has the same rating. In some lamps the switch can be separated from the socket; in others the two are integral. Before buying a new switch, open the housing and see just what it is you need.

To take apart the switch housing, look for small screws which hold it together. Some housings are held together by friction and can be pulled apart. Notice in the figure of the three-way lamp that one wire from the power line goes directly to the bulb filaments and does not go through the switch. However, in the lamp, you will find all the wires in a bunch and you must separate them to determine which is the *common* wire. You usually will leave the common wire undisturbed and remove only the wires actually connected to the switch. Note how they are screwed on to the switch terminals and replace them in the same manner on the new switch.

2-3. Timers

Large automatic electrical appliances must be controlled so that each function is performed in the proper sequence. In a dishwasher, for instance, first a pump sucks out any water which has accumulated, then a valve opens to let in hot water. When the water level is correct, the valve closes, and the motor turns on and sprays the soap and water over the dishes. The motor subsequently stops, and the water is pumped out. Clean water enters and the process is repeated several times. Finally a heating element is activated to dry the dishes. Each function is turned on and off by an electric current, but a

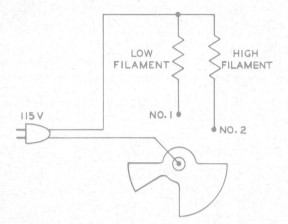

Fig. 2-6. Three-way lamp.

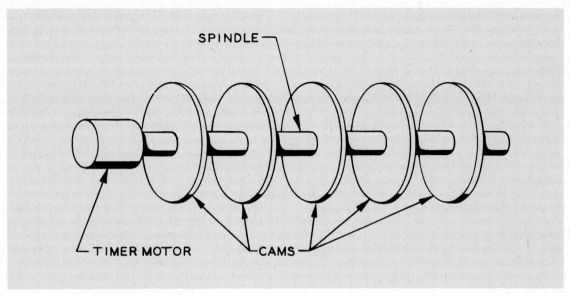

Fig. 2-7. Timer and cams.

master control is needed to open and close the switches at the proper time in the cycle. The master control is called a *timer*.

The heart of the timer is simply an electric clock motor which runs at a constant speed. A set of gears reduces the speed so that the motor rotates a spindle at the rate of one revolution for one complete operation of the appliance. This is similar to the gear train in an electric clock which steps down the motor speed so that the minute hand makes one revolution per hour. The exact gear ratio

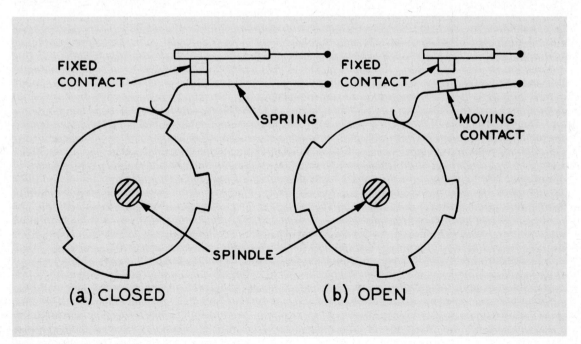

Fig. 2-8. Cam and switch.

These are some of the tools the home owner should have for making routine repairs about his house.

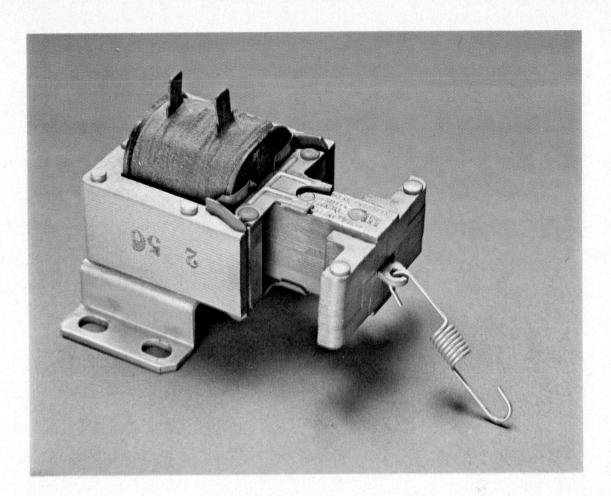

varies with the appliance. If the complete wash-rinse-dry cycle should take 55 minutes, for example, the gear ratio will allow the spindle to make one revolution in 55 minutes.

Mounted on the spindle and firmly attached to it are several cams, as shown in Figure 2-7. These cams have projections on them to actuate switches, and each cam controls a separate function of the appliance. Thus, one turns the motor on and off, another controls the valve for water flow, etc. A typical cam and the switch it controls are shown in Figure 2-8. A spring rides on the edge of the cam and when it is raised by a projection, it closes a pair of contacts, as shown in (a). In (b), the spring is between projections and the contacts are open. The number of projections on each cam and the length of each projection

correspond to the number of times the switch must be closed and the length of time it remains closed.

If the cam turned at a constant speed, the switches would open and close slowly, and the current flowing at the moment of opening would cause sparking which would pit the contacts, possibly welding them shut or preventing them from making good contact. To avoid this difficulty the spindle is spring-loaded in such a manner that it moves in short impulses instead of turning smoothly. Each impulse is rapid, but with the pauses between, the total time of rotation is still held to one complete cycle of the appliance.

This type of timer consisting of cams and switches is the most common type, but there are others. A somewhat different kind of timer has a long screw in place of the spindle. A carriage is threaded on the screw and moves horizontally from one end to the other as the screw rotates. Metal fingers on the carriage make appropriate contacts, closing circuits, as it moves along.

A timer motor rarely needs servicing, but if it fails, it should be replaced by an identical part. Before removing the old timer motor, however, make sure that it is receiving voltage at its terminals. If there is voltage, but the timer motor does not run, see if there are mechanical restrictions, such as stuck bearings in the gear box, before deciding the motor is bad.

A more common fault in a timer is a pair of switch contacts which are welded shut from arcing. The switch levers should move easily when pushed with the fingers and should spring open when released. If contacts are stuck together, pry them apart, being careful not to bend the springs, and clean them with sandpaper. Contacts also may be pitted and burned so badly that they fail to make an electrical connection. Again, they should be sandpapered smooth. Never use emery paper, since the grit is conductive and can cause a short circuit if particles fall in the wrong place.

If a cam gets loose on the motor spindle, you can tighten it, but first you must find out where in the cycle the particular switch closes. Some manufacturers supply this information in the instruction sheets in the form of a timetable for the different functions. You can also get the information from an authorized service dealer. If the cam is twisted on the spindle, the switch will open and close at the wrong times.

2-4. Solenoids

A *solenoid* is an electromagnet. Basically it consists of a coil of wire and an iron bar which is free to move through the coil. A spring holds the bar outside of the coil when the circuit is disconnected, as shown in Figure 2-9. When the coil is energized, that is, when a voltage is

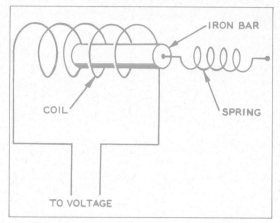

Fig. 2-9. Solenoid.

applied, current flows and makes the coil a magnet. The iron bar is pulled into the center of the coil. When the voltage is disconnected, the spring pulls the bar out again. The bar can be attached to a mechanical linkage to shift gears, tighten a drive belt, open or close a valve, start or stop an agitator in a washer, or perform other mechanical functions.

An example of a water valve, using a solenoid, is shown in Figure 2-10. When the coil is energized, the pointed iron rod is pulled up, as indicated, and water flows through the pipe to the tub. When the electricity is turned off, the bar is pushed down and blocks the pipe, stopping the flow of water. The coil itself is usually a separate entity which is slid over the outside of the valve and is easily removable for testing or replacing.

A *relay* is a solenoid which controls a high current or voltage. For example, some of the functions in a dishwasher, such as drying, draw currents which are too great for the small contacts in the timer. The small contacts close a circuit to energize a solenoid, and the solenoid closes a pair of heavier contacts to start a larger current. The basic relay circuit is shown in Figure 2-11. The solenoid coil is wound permanently on an iron core to make a strong electromagnet.

When the switch is closed, current flows through the coil, making it a magnet. It attracts the movable soft iron bar holding one heavy contact. Since one end of this bar is fixed on a pivot point, the bar rotates on this pivot, and in doing so, the heavy contact on

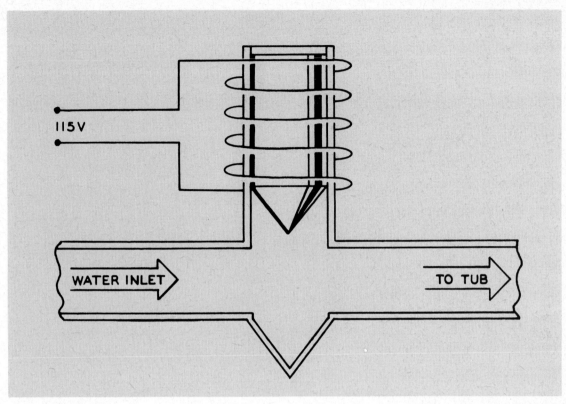

Fig. 2-10 Water valve.

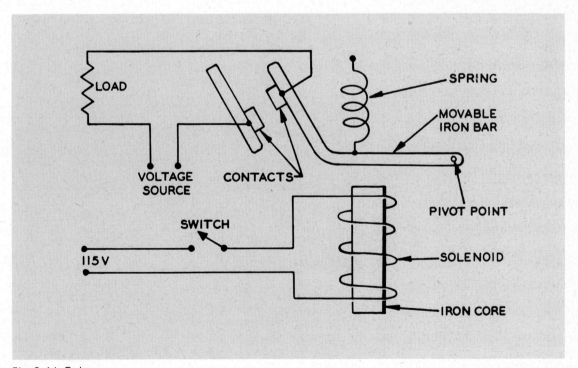

Fig. 2-11. Relay.

the bar is brought against a *fixed* contact closing the electric circuit to the load. The voltage source can be a heavy-duty 240-volt line or a battery or whatever is required. When the switch is opened, the solenoid no longer attracts the movable arm, and the spring pulls the arm away and thus separates the contacts. The arm is shown in the open position in Figure 2-11

The most common trouble in a solenoid or relay is an open coil or a grounded coil. If a solenoid fails to operate, make a continuity test at its terminals. You should read a small resistance on an ohmmeter, the resistance of the coil. If you use a test lamp, the light should glow, indicating continuity. Also check continuity between each of the solenoid terminals and the case. Here you should have no continuity unless the coil is grounded. If the coil is open or grounded, it can usually be replaced easily. In some small relays, it may be cheaper to replace the whole relay.

Check relay contacts for continuity when they are closed and, if necessary, sandpaper them smooth. Many troubles with relays and solenoids are mechanical. The pivot point in the relay may need a drop of oil. Valves which are operated by solenoids may be stuck. In general, if the solenoid coil is all right, the trouble is not in the solenoid, but is probably a mechanical problem in the part that is operated by the solenoid.

Types of heating elements.

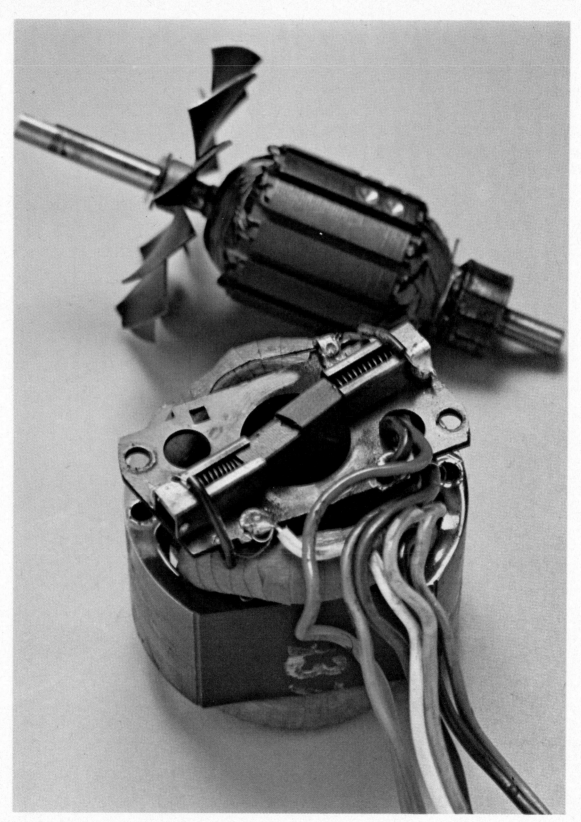

Interior of a motor (See Chapter 1).

Heating Elements and Thermostats

The simplest electrical appliances and thus the easiest to service are those that produce heat. All contain some sort of heating element which gets hot when an electrical current passes through it. If an appliance is *automatic*, that is, if it maintains a certain temperature without manual adjustment, then it usually contains a thermostat as well.

3-1. Heating Elements

Heat is produced when an electric current is passed through a resistance. The power dissipated in the resistance, as discussed in Chapter 1 of Volume 4, is the product of voltage and current. That is,

Watts Volts × Amperes.

The resistance of a wire is expressed *ohms*, which is the ratio of volts to amperes. That is,

Ohms Volts/Amperes

Amperes Volts/Ohms

These formulas are usually expressed as follows:

P E X I

R E/I

I E/R

where P is the power in watts, E is the electromotive force (EMF) in volts, I is the intensity of the current in amperes, and R is the resistance in ohms.

For example, assume the line voltage is 115 volts and the current through a toaster is 10 amperes. Then,

Power 115 × 10 1150 watts

Resistance 115/10 11.5 ohms

It should be noted than when a toaster is turned on, current flows through the line cord and the heating element in series, as depicted in Figure 3-1. The line cord has a very small resistance, much less than 0.05 ohm. The heating element for the 1150-watt toaster has a resistance of 11.5 ohms, as shown in the example above. The two resistances are said to be in *series* since the same current flows through both. The 115 volts divides across the resistances in proportion to the values of resistances. That is, since 11.5 is more than 200 times 0.05, we would expect the voltage across the heating element to be about 200 times the voltage across the line cord. Thus, we might expect to measure about half a volt across the line cord and about 114.5 volts across the heating element. The figure of 115 volts for the line voltage is only an approxima-

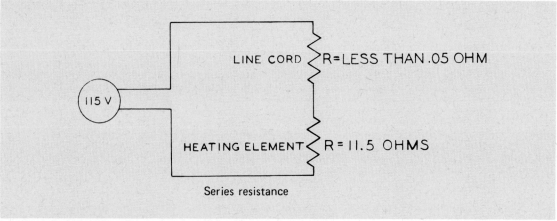

Fig. 3-1. Series resistance.

tion, since the line usually varies from 110 to 120 volts, depending on the total load at the power station. Since the line cord resistance is so small compared to that of the heating element, we can neglect it and assume that the 115-volt line potential is entirely across the heating element.

Note, however, that if the line cord has some strands broken so that its resistance increases to only one ohm, the new division of voltage could be detrimental. Now the one-ohm line cord is in series with the 11.5-ohm heating element, and the total resistance across the line is 12.5 ohms. The current is then

115 volts / 12.5 ohms 9.2 amperes

The voltage divides according to the resistance. That is, the voltage across the line cord is 1/12.5 of the total voltage, and the voltage across the heating element is 11.5/12.5 of the total voltage. This means that the voltage across the heating elements is only 105.8 volts. The power dissipated in the heating element is then

105.8 volts × 9.2 amperes 973.36 watts

This is about 85 per cent of the rated power of 1150 watts and means that the toaster would produce only 85 per cent of the heat it should. This is another reason why it is very important to make sure line cords are in good condition.

The resistance of the heating element determines the power that will be used. Thus, if you want a 1000-watt element, you know that the current through it should be 1000/115 or 8.70 amperes (current watts/volts). Since resistance is volts divided by amperes, you know that the resistance should be 115/8.70 or 13.2 ohms. Therefore, you need a 13.2-ohm heating element to dissipate 1000 watts. Values of current and resistance for several power ratings are given in Table 3-1.

Since ordinary copper wire has very low resistance, it is used in line cords and other parts of the circuit where heat is *not* wanted.

Table 3-1. Current and Resistance as a Function of Power

(115-volt line)

Power (watts)	Current (amps)	Resistance (ohms)
500	4.35	26.4
750	6.52	17.6
1000	8.70	13.2
1250	10.87	10.9
1500	13.04	8.82

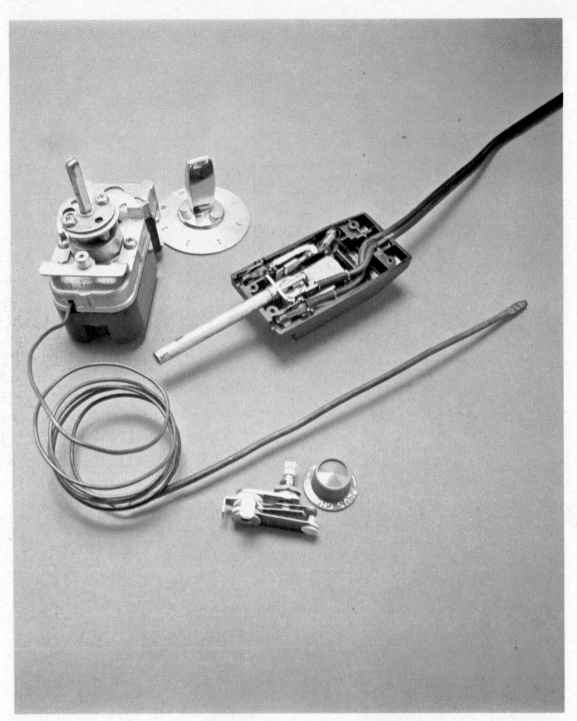

Types of thermostat.

Special materials with much higher values of resistance are used where it is necessary to produce heat or light. Most of these materials produce both heat and light when power is dissipated in them, but they are chosen for the particular application on the basis of which they produce better. Thus for example, tungsten is used for light bulb filaments because it produces light well, although it also gets hot. Similarly, a heater element in a toaster glows when it gets hot, but heat is its most important output. From Table 3-1, a 500-watt unit should have a resistance of 26.4 ohms. This is true whether it is a 500-watt bulb giving off light or a 500-watt heater producing heat.

The most common material used in heating elements is *nichrome*, an alloy of nickel and chromium. For the same size wire, nichrome has a resistance more than 1000 times that of copper. In toasters, flat nichrome wire or ribbon is used as the heating element since it can be maintained at a fixed distance from the bread to be toasted. Nichrome ribbon has a resistance of about 1/20 ohm per inch, which means a length 20 inches long has only one

ohm resistance. As you can see in Table 3-1, the resistance for a 1000-watt toaster must be 13.2 ohms, and since each ohm requires 20 inches of ribbon, the total length of the heating elements must be 264 inches, or 22 feet. There are usually four elements in a toaster, one on each side of each piece of bread, so that each of the elements must be one-fourth of 22 feet, or 5.5 feet. This length is accommodated in the toaster by winding the ribbon back and forth on mica supports. Figure 3-2 is a cutaway sketch of a toaster, showing the nichrome ribbon heating elements.

In order to reduce the length of the heating element, the nichrome wire is frequently coiled on a one-quarter inch diameter. This is the form of heating element used in hotplates, as shown in Figure 3-3, as well as inside waffle irons, grills and similar appliances. Note that the heating element is one continuous coil from point A to point B. Coiled nichrome wire has a resistance of approximately 0.4 ohm per inch of coil. Using this value, a 1000-watt appliance would need a

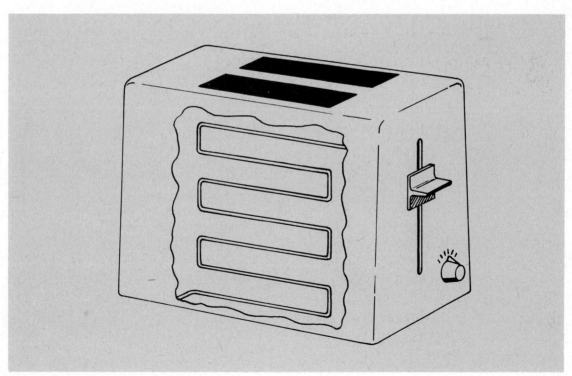

Fig. 3-2. Toaster heating element.

Close-up of coiled nichrome heating element.

heating element 33 inches long as compared to the 264 inches of nichrome ribbon. Lengths of heating elements for both nichrome ribbon and coiled nichrome wire are given in Table 3-2. Note that as the length of the element is decreased, the resistance also is decreased, and the power is increased. This means that if a heating element broke and you tried to make a temporary repair by discarding the shorter piece and stretching the longer piece to join it to its terminal, you would be using a lower resistance and would thus have more current and more power. You would actually get more heat, but should be careful not to exceed the current rating or power rating of other parts of the circuit.

Both nichrome ribbon and coiled wire are

Power (watts)	Resistance (ohms)	Nichrome ribbon 0.05 ohm per inch (inches)	Coiled nichrome wire 0.4 ohm per inch (inches)
500			
750	26.4	528	66
1000	17.6	352	44
1250	13.2	264	33
1500	10.9	218	27
	8.82	176	22

Table 3-2. Length of Nichrome Heating Elements.

Fig. 3-3. Coiled nichrome heater.

the switch is in the grounded side, then the heating element is connected directly to the hot side of the line. If you touch this and are grounded at the same time, you will get an electric shock just as if you touched both sides of the line simultaneously. If you must use a knife to pry out a piece of bread stuck in a toaster, make sure you pull the wall plug first.

Heating elements that must be exposed, as for example on the top of a range, are usually made of coiled nichrome wire completely enclosed in a metal sheath, from which the nichrome wire is insulated electrically. The trade name for one such type of heating element is Calrod, and this name is usually applied indiscriminately to all sheathed heating elements. Calrod heaters as used on top of a kitchen range are shown in Figure 3-5. The metal sheath also protects the nichrome wire from oxidizing.

Heating elements are usually connected to their terminals with bolts. Solder could not be used, since it would melt from the intense heat. If a heating element breaks, you should replace it with one of the same electrical rating and physical size. Sometimes a temporary repair can be made by fastening the two broken ends together with a bolt and nut, as shown in Figure 3-6. Each of the two wires is wound around a small bolt, and flat washers

used in heating elements which are enclosed so that the user will not be apt to touch the elements. Note that there is a potential danger of shock from touching a heating element and ground at the same time whenever the element is connected to the line, even if it is not turned on. This is explained in Figure 3-4. The switch in the circuit is usually in one side of the line only. (In a toaster it is actuated by pushing down the plunger.) If the line cord is plugged into a wall outlet so that

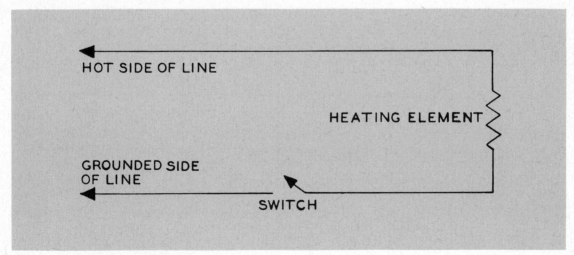

Fig. 3-4. Potential shock hazard.

Fig. 3-5. Calrod heating elements.

are used to maintain even pressure. The bolt must be insulated from the rest of the circuit and from anything that could be in contact with the user. Note that this is strictly a *temporary* solution, and the heating element should be replaced as soon as possible.

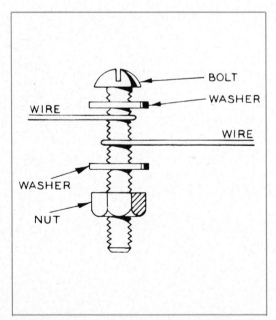

Fig. 3-6. Quick repair of heating element.

3-2. Thermostats

A *thermostat* is a switch which is controlled by heat. The most common type used in electrical appliances is a *bimetallic* thermostat which consists of two strips of different metals firmly fastened together, as shown in Figure 3-7. One of the metals has a high rate of thermal expansion, while the other has low thermal expansion. When cold, the combined strip is straight, as shown in (a). When the unit is heated, the two metals expand at different rates, and the alloy with a high coefficient of thermal expansion becomes longer than the other. This results in the unit bending, as shown in (b). When the thermostat cools, it returns to the configuration in (a).

A thermostat can be used to make a contact or to break one. In Figure 3-7, two terminals, A and B, are shown, one on each side of the thermostat. Two contacts are also shown on the thermostat itself, C and D. When the thermostat is cold, as in (a), contact D touches terminal B, and connects B to the circuit (represented by the wire screwed to the left end of the thermostat). When heat is

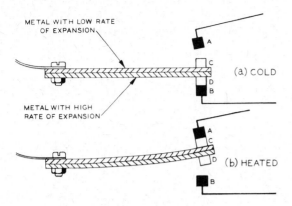

METAL WITH LOW RATE
OF EXPANSION

(a) COLD

METAL WITH HIGH
RATE OF EXPANSION

(b) HEATED

Fig. 3-7. Thermostat.

applied, the thermostat bends, breaking this connection, and now contact C connects terminal A to the circuit, as shown in (b). In practice, only one contact and one terminal are usually used. If the thermostat is used to break a circuit when it gets hot, then only contact D and terminal B are used. When cold, the thermostat connects these two together, and when hot, it breaks this connection. On the other hand, if the thermostat is needed to turn on a circuit when it gets hot, then only A and C are used.

The temperature at which the thermostat makes or breaks a connection can be controlled very accurately by the spacing between the contact and the terminal. In some appliances, such as waffle irons and coffee percolators, it is necessary to bring the heat up to a certain specified temperature and to maintain it there. At the start, the thermostat closes the circuit since it is designed to make a connection when cold. As the temperature increases and the thermostat bends, it finally reaches a

Bimetallic thermostat.

point where it breaks the circuit by separating contact D from terminal B. Since there is some springiness in the metals, contact B can be pushed up so that the break occurs at the correct temperature. In many appliances, this temperature may be varied by turning a knob or moving a lever which moves contact B up or down. On a waffle iron or toaster, this knob may have a pointer which is set on a variable scale from ''light'' to ''dark''. In any case, as soon as the contact is broken, the heating element in the appliance no longer operates, and the temperature starts to decrease. Now the thermostat bends back and closes the connection again, and the whole cycle is repeated. By continually making and breaking the contact in this way, the thermostat holds the temperature in the appliance within narrow limits. By suitable choice of alloys it is possible to design a thermostat to maintain a constant temperature in an appliance within one degree, although in practice five or ten degrees of variation is acceptable.

Thermostats are used in most appliances using heat, such as electric blankets, toasters, waffle irons, irons and others. Wherever a variable temperature is desired, this is set by controlling the spacing of the contacts. The operating temperature can be changed over a wide range in this manner.

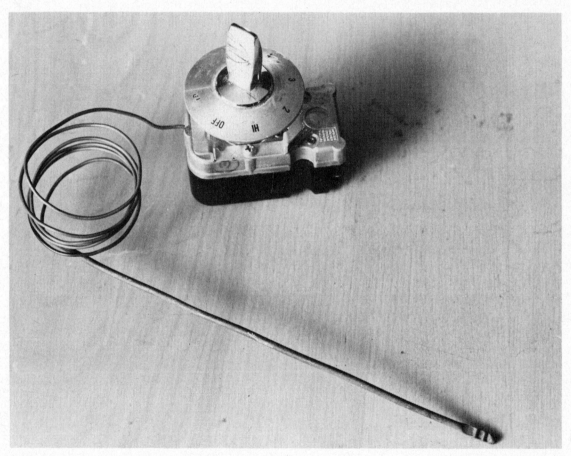

Thermostat used for burner switch on electric stove.

Large Appliances

Large electrical appliances found in most households include refrigerators, dishwashers, clothes washers and dryers, and garbage disposers. To be able to fix every breakdown of every one of these appliances, the owner would need some special tools, as well as a large assortment of the more common household tools. When a special tool is called for, it rarely pays to invest in one, since the same trouble may never arise again. To add to the problem, most of these large appliances are heavy or awkward to get at, and frequently an extra pair of hands or some sort of moving apparatus is needed. Nevertheless, there are many faults you can correct, and if you can recognize these, you may be able to save the price of a few service calls.

4-1. Garbage Disposers

Garbage disposers grind or shred organic material into particles small enough to be washed down the drain. These materials include bones, vegetable matter including fibrous materials, and practically all food waste. Garbage disposers are *not* supposed to grind glass, string, bottle tops, silverware or broken dishes. With proper use, a garbage disposer gives trouble-free service for many years.

Because it must grind heavy bones as well as soft materials, a garbage disposer must be both ruggedly built and able to operate under varying load conditions. Usually it uses a capacitor-start motor which has a moderately good starting torque, although some less expensive models use split-phase motors. The motor turns an impeller and a flywheel; there are no other moving parts.

The operation of a garbage disposer is illustrated in Figures 4-1 and 4-2. Basically, the machine consists of two main housings

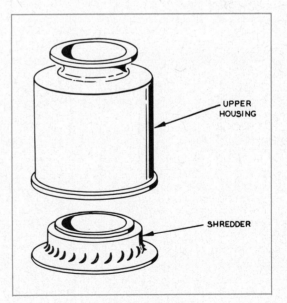

Fig. 4-1. Upper housing of disposer.

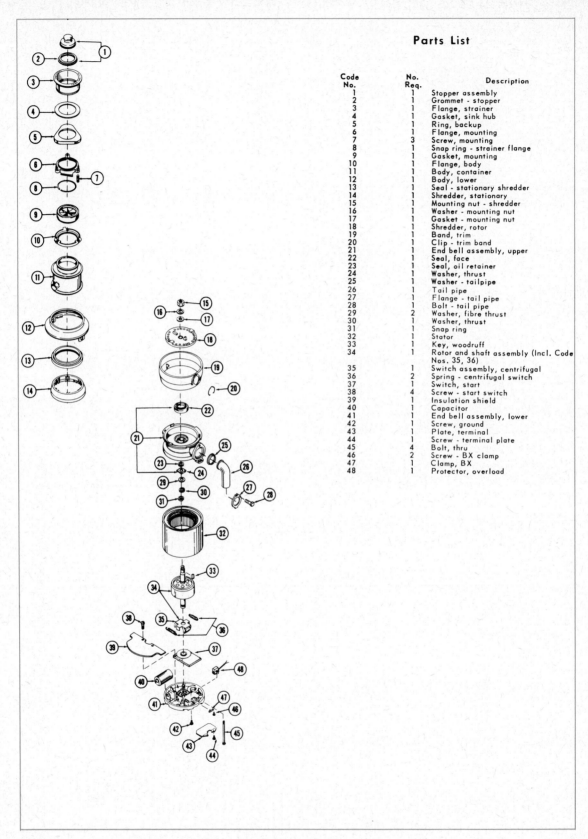

Parts List

Code No.	No. Req.	Description
1	1	Stopper assembly
2	1	Grommet - stopper
3	1	Flange, strainer
4	1	Gasket, sink hub
5	1	Ring, backup
6	1	Flange, mounting
7	3	Screw, mounting
8	1	Snap ring - strainer flange
9	1	Gasket, mounting
10	1	Flange, body
11	1	Body, container
12	1	Body, lower
13	1	Seal - stationary shredder
14	1	Shredder, stationary
15	1	Mounting nut - shredder
16	1	Washer - mounting nut
17	1	Gasket - mounting nut
18	1	Shredder, rotor
19	1	Band, trim
20	1	Clip - trim band
21	1	End bell assembly, upper
22	1	Seal, face
23	1	Seal, oil retainer
24	1	Washer, thrust
25	1	Washer - tailpipe
26	1	Tail pipe
27	1	Flange - tail pipe
28	1	Bolt - tail pipe
29	2	Washer, fibre thrust
30	1	Washer, thrust
31	1	Snap ring
32	1	Stator
33	1	Key, woodruff
34	1	Rotor and shaft assembly (Incl. Code Nos. 35, 36)
35	1	Switch assembly, centrifugal
36	2	Spring - centrifugal switch
37	1	Switch, start
38	4	Screw - start switch
39	1	Insulation shield
40	1	Capacitor
41	1	End bell assembly, lower
42	1	Screw, ground
43	1	Plate, terminal
44	1	Screw - terminal plate
45	4	Bolt, thru
46	2	Screw - BX clamp
47	1	Clamp, BX
48	1	Protector, overload

Exploded view and parts list of a continuous feed disposer (Courtesy In-sink-motor).

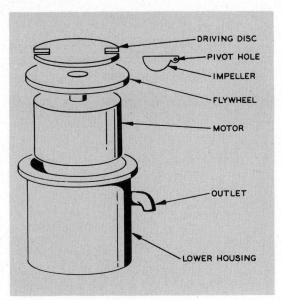

Fig. 4-2. Lower housing of disposer.

shredded to bits. Cold water should be running while the disposer is in operation. This water washes the shredded garbage down the drain by way of the outlet shown in Figure 4-2. A trap must be used in the outlet line. Note that cold water is used because it hardens any fat in the garbage, and the disposer is then able to grind the fat into tiny particles. If hot water were used, the fat would melt and flow into the drain where it could solidify when it cooled, causing a blockage.

A garbage disposer must have at least one switch to turn it on. Some units have a simple wall switch, similar to a light switch. Others have a switch which is controlled by the sink stopper so that it is impossible for the user to stick his hand in the disposer while it is on. An additional switch may be placed in the cold

which are held together by bolts and nuts with a suitable gasket between the housings to prevent water leaks. The upper housing, shown in Figure 4-1, contains a chamber in which the garbage to be shredded is placed. At the bottom of the chamber there is a fixed shredder which is simply a metal ring with shredding teeth on its inside circumference. The bottom of the garbage chamber is open, so that the waste material rests on a disc at the top of the lower chamber. This disc is shown at the top of Figure 4-2, which illustrates the bottom housing of the disposer. This housing contains the motor which rotates the driving disc and the flywheel. These parts are shown separated in the figure, but they are mounted on the motor shaft. Seals are required around the rotor shaft to prevent water from leaking into the motor. Washers, gaskets and other hardware are not shown in the figures.

In operation, the motor rotates the disc, causing the garbage to fly outward by centrifugal force. Near the periphery of the disc are two or more impellers, free to move on pivots, as shown in Figure 4-2. These lie flat when the motor is at rest, but fly outward in motion so that they are close to the grinding teeth of the shredder. The garbage falls between the grinding teeth and the impellers and is

Garbage disposer.

Troubleshooting Chart   FOOD WASTE DISPOSERS

PROBLEM	POSSIBLE CAUSE	CORRECTIVE ACTION
Leaks at sink flange.	Loose mounting screws.	Tighten flange screws.
Leaks at drain gasket.	Loose flange. Improper gasket.	Tighten screws. Replace.
Leaks between chamber and sink flange.	Loose mounting. Defective gasket.	Replace. Tighten nuts.
Abnormal noise and/or vibration.	Undisposable matter in flange gasket or tailpipe gasket improperly placed. Broken impeller vane. Motor bearings may be damaged.	Clean out chamber. Replace. Replace impeller. Replace bearings.
Erratic operation.	Loose wiring, switch, motor, or power connection.	Locate and reconnect.
Slow grinding.	Undisposable matter. Damaged impeller. Dull shredder. Insufficient water flow.	Remove. Replace. Replace. Minimum 2 gallons per minute.
Slow drain-out.	Partially clogged drain. Clogged shredder teeth.	Check plumbing. Remove and clean.
Jammed.	Something stuck between impeller and shredder.	Move impeller backward until free.
Won't stop.	Defective switch. Short in wiring. Incorrect wiring.	Replace. Clear short and insulate. Reconnect properly.
Won't run.	Overload protector has tripped. Blown fuse. Defective switch. Burned-out motor winding. Open or shorted wiring. Inoperative centrifugal switch in motor.	Wait for motor to cool; reset. Replace. Replace. Replace stator. Repair. Replace.
Cover doesn't control.	Defective switch. Broken guide tab. Worn cover.	Replace. Replace housing. Replace.

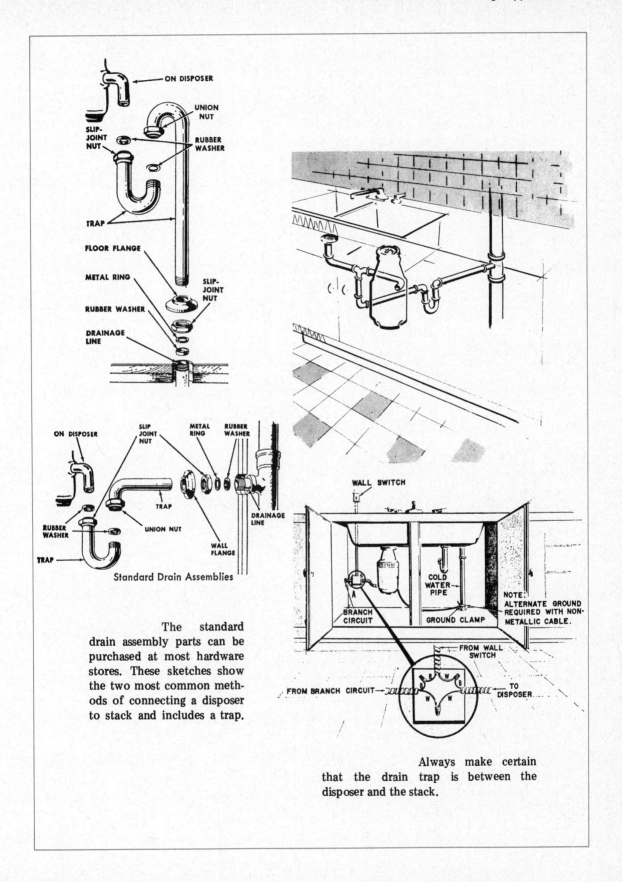

ON DISPOSER

UNION NUT

SLIP-JOINT NUT

RUBBER WASHER

TRAP

FLOOR FLANGE

METAL RING

SLIP-JOINT NUT

RUBBER WASHER

DRAINAGE LINE

ON DISPOSER

SLIP JOINT NUT

METAL RING

RUBBER WASHER

TRAP

RUBBER WASHER

UNION NUT

TRAP

DRAINAGE LINE

WALL FLANGE

Standard Drain Assemblies

WALL SWITCH

COLD WATER PIPE

BRANCH CIRCUIT

GROUND CLAMP

NOTE: ALTERNATE GROUND REQUIRED WITH NON-METALLIC CABLE.

FROM WALL SWITCH

FROM BRANCH CIRCUIT

TO DISPOSER

The standard drain assembly parts can be purchased at most hardware stores. These sketches show the two most common methods of connecting a disposer to stack and includes a trap.

Always make certain that the drain trap is between the disposer and the stack.

water line and is actuated only when the cold water is flowing. If this additional switch is used, both switches must be on before the disposer will operate. The motor also has a built-in overload switch which shuts off the electricity when the disposer has too heavy a load. In most models, the overload switch can be returned to the on position by pushing a button at the bottom of the unit.

Garbage disposers should give at least five years of trouble-free operation if used properly. Eventually the seals may crack, letting water into the motor, or parts may wear out. However, occasionally a disposer may fail to operate because it is overloaded or stuck. This is easily corrected, and, in fact, is to be expected, since food particles sometimes lodge in such a position as to jam the machine.

If the disposer does not turn on, but the motor hums, it is usually a sign that particles of garbage have jammed the unit. This sometimes happens when the disposer has been shut off too soon, leaving some food particles unshredded. Then the next time you try to use the disposer, these particles cause trouble by being wedged between the impellers or driving disc and the stationary part of the unit. With all switches off and with the disposer unplugged, if possible, reach into the opening and try to turn the driving disc. You can usually take hold of one of the impellers and use it as a handle to move the disc. *It should turn easily.* If food is jamming the disposer, the disc cannot be turned. Usually you can free the jam by turning the disc backwards. Manufacturers supply a special wrench, in the shape of a large Z, to do this. It is inserted in the opening and pushed against the impellers to free the food particles. You can also use a piece of wood. In some newer models, there is a special reversing switch, so that the disposer can be freed electrically.

If the jam cannot be cleared by using a wrench or piece of wood, it will be necessary to remove the bottom housing from the disposer. The upper housing is fastened to the sink. The bottom housing may be removed by unscrewing the six or more screws which hold the two housings together. This is not difficult, but it might be awkward. There is not much room for a regular screwdriver, so it may be necessary to use an offset screwdriver. In addition, the bottom housing is heavy, and you must be careful not to let it drop when the screws are removed. When the two halves are separated, it is a simple matter to remove the particles of food or bone causing the obstruction.

Jamming can also be caused by frozen bearings. You can determine this when the lower housing is removed. It is possible to replace bearings for the flywheel, and on some disposers the motor bearings are also easily accessible. On others it is necessary to tear the whole motor apart. Fortunately, bearings last many years, and when they do fail, it may be well to consider replacing the disposer rather than just the bearings. After many years the shredding teeth are worn, and the disposer doesn't operate as well as it did when it was new.

If the disposer shows no sign of life, make sure the overload button is reset and all switches are in order. Switches should be checked for continuity in the on position. *Make sure the disposer is disconnected from the power line when making these tests.* If the switch in the cold water line fails, you can bypass it temporarily. You must then remember to turn on the cold water when using the disposer.

The switch in the sink opening may show continuity, but may be at fault because it has moved. Check to make sure that it is turned on when the stopper is placed in the proper position, and if not, adjust it so that it does and tighten the bolts that hold it in place.

If a disposer won't turn off, the trouble is a stuck switch. Repair or replace it as necessary. If the disposer runs and shuts off by itself, it may be overloaded, or the overload switch may be defective. To replace a bad overload relay or switch, it is necessary to remove the lower housing and take out the motor.

If a disposer takes too long to grind the garbage, the impellers may be stuck so that they are not pushing the food against the shredder. Reach into the opening and try moving the impellers. You should be able to move them freely with your fingers. If one is stuck, you can pry it loose with a small screwdriver or the back of a teaspoon. On rare occasions you might have to remove the lower housing to remove the food particles. If the impellers are free, slow shredding is probably due to worn teeth in the shredder. Replace the shredder. To do this, remove the lower housing and you can then reach the shredder in the top.

If a disposer leaks, it is usually a sign of worn or broken bearings. These can be replaced, but as a rule this doesn't happen until the disposer is old, and it is usually better to replace the whole disposer at this time.

4-2. Refrigerators

All substances exist in one of three physical states, namely, *solid, liquid* or *gas*. It is possible for a substance to change its state by the addition or removal of heat from the substance. When a substance changes from solid to liquid, from liquid to gas, or directly from solid to gas, heat *must be added* to the substance. When the change of state is in the other direction, that is, from gas to liquid, from liquid to solid, or from gas to solid, heat

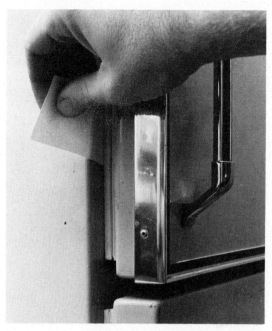

To check the door gaskets of a refrigerator, insert a piece of paper between the door and the frame, and pull: there should be some resistance.

must be *removed* from the substance. In practice, this means that, for example, when a liquid evaporates (changing from liquid to gaseous state), it must absorb heat from the surrounding medium. As an experiment, try moistening your hand with alcohol. As the alcohol evaporates, it draws heat from your skin, and your hand feels cool.

Anything which accelerates the change of state increases the addition or removal of heat. Thus, in the alcohol experiment above, if you fan your hand or blow on it, you will cause the alcohol to evaporate more rapidly , and your hand will feel cooler than it would without the additional air motion. Another method of increasing evaporation is by lowering the pressure. Water boils at 212F at sea level, but on top of a high mountain, the boiling point will be several degrees less. The air pressure on the surface of the liquid is less on the mountain top, so that boiling takes place at a lower temperature. In a vacuum, water would boil even at its freezing point of 32F. Below the boiling point, evaporation is increased as the pressure on the surface of the liquid is decreased, so that on a mountain top water evaporates more rapidly than it does at sea

Refrigerator with freezer unit at top.

level. Conversely, if the pressure is increased, evaporation is slowed.

These two principles are the basis of operation of a refrigerator. Before electric refrigerators were invented, people used iceboxes. A piece of ice placed in the ice chamber slowly melted. The change of state from solid water to liquid water required heat, and this heat was drawn from the surrounding air and from the food in the icebox. If the ice were wrapped in paper to make it last longer, the rate of melting would be decreased, and the food would not get as cool.

In electric refrigerators, a special *refrigerant* is used. This is a substance which boils at a low temperature, much below the freezing temperature of water. Another desirable characteristic of a good refrigerant is that it must change to liquid when the pressure is increased. There are many substances which can be used as refrigerants, all with long chemical names. They are usually referred to by their trade names. Typically, *Freon* boils at 22° below zero on the Fahrenheit scale, and *Refrigerant-22* boils below 40° below zero. In comparison, the boiling point of water is 212° F, and its *freezing point* is 32F. Thus, Freon, for example, boils at 54° less than the temperature at which water freezes.

A simple refrigerator could be made as shown in Figure 4-3. A jug of refrigerant is placed in one part of an insulated box with the opening of the jug leading to the outside. As the refrigerant boils or evaporates, it draws heat from the rest of the box, cooling the food there. You should recognize that this is essentially the same as the old icebox, with a liquid refrigerant replacing the block of ice. Both depend on a change of state for cooling

Occasionally vacuum the condenser cords on the back of your refrigerator.

the food. The refrigerator of Figure 4-3 has several drawbacks, however. Refrigerants are expensive, and it would be necessary to keep replenishing the material that boiled away. Also, good chemical refrigerants may be harmful to health, and their vapors might be dangerous floating around the room. If a refrigerant like Freon is used, the temperature of the food would fall well below the freezing point of water, which is much too low for proper storage.

A practical refrigerator, then, must have some means of catching the escaping vapor or gas and converting it back to liquid so that it is not necessary to add more refrigerant. Also, there must be some means of controlling the inside temperature by controlling the evaporation of the liquid. The basic circuit of a practical refrigerator is shown in Figure 4-4. The refrigerant is contained in a closed system of pipes and is moved through the system by a compressor. The compressor is simply a pump driven by an electric motor. It sucks in warm vapor at the left in the figure and pushes it into a condenser, which is simply a length of pipe passing back and forth so that the whole length can be contained at the back of the case of the refrigerator. Note

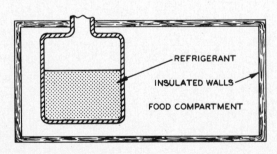

REFRIGERANT

INSULATED WALLS

FOOD COMPARTMENT

Fig. 4-3. Simple refrigerator.

Do not pile things on top of a refrigerator as it works better when air circulates freely around it.

Troubleshooting Chart   REFRIGERATORS

PROBLEM	POSSIBLE CAUSE	CORRECTIVE ACTION
Compressor will not run.	No power at outlet.	Replace fuse.
	Loose electrical connection.	Repair.
	Over heated compressor.	Clean condenser, check fan, clearance around cabinet.
	Thermostat stuck open or turned off.	Replace if needed.
	Defective relay, overload or compressor.	Use starting cord to check compressor. If it runs, replace defective overload on relay. If not replace compressor.
Compressor runs but no refrigeration.	Restriction in system, moisture or permanent.	Repair as outlined in refrigeration chapter.
	Low charge.	Locate and repair refrigerant leak. Evacuate and recharge.
	Inefficient compressor.	Replace as outlined.
Compressor kicks out on overload.	Abnormal usage or high room temperature.	
	High or low voltage, outside 10% tolerance.	
	Impeded air circulation.	Clean condenser, check fan if used. Provide space for air circulation.
	Defective capacitor.	Replace capacitor.
	Defective start relay.	Replace start relay.
	Defective overload protector.	Replace overload protector.
	Defective motor winding.	Replace compressor.
Refrigerator too cold.	Thermostat set too cold, or contacts stuck.	Check setting. Replace if contacts are bad.
	Abnormal location.	
Refrigerator too warm.	Restricted air circulation.	Clean condenser, check fan and cabinet clearance.
	Abnormal usage, location, or high room temperature.	
	Poor door seal.	Adjust door closing.
	Light switch stuck on.	Replace switch.
	Defrost heaters on all the time.	Check defrost timer and thermostat. Readjust or replace if defective.
	Operating thermostat set too warm, or has defective sensor.	
	Defective compressor.	Replace compressor.
	Cooling fan not running.	Check for voltage at fan. If none, check wiring. If O.K., check fan.
	Restricted air duct.	Clear air duct.
	Leaking air duct.	Reseal joints.
	Excessive frost on evaporator.	Defrost evaporator.
	Open defrost heater will allow ice to restrict duct.	Replace heater.

Troubleshooting Chart   REFRIGERATORS (continued)

PROBLEM	POSSIBLE CAUSE	CORRECTIVE ACTION
Noisy unit.	Loose parts.	Tighten loose parts.
	Tubing rattle.	Move tubing.
	Abnormal fan noise.	Check fan blade security, motor mounts and blade clearance from shroud.
	Abnormal compressor noise.	Check mounting pads for clearance. If noise is internal, replace compressor.
Sweating outside of cabinet.	Damp or confined location.	If possible relocate refrigerator.
	Low side tubing too close to cabinet shell.	Reposition tubing.
	Mullion or stile heaters may be open.	Replace defective heaters.
	Void in insulation.	Replace insulation.
	Wet insulation.	Correct cause, replace insulation.
Sweating inside of cabinet.	Abnormal usage.	cover foods and defrost as needed.
	Poor door seal.	Make necessary door adjustments.
	Wet insulation due to poor cabinet seal.	Remove liner, replace insulation and reseal cabinet.
Incomplete defrosting or high temperature during defrost cycle.	Defrost thermostat may be loose or defective.	Remount or replace thermostat.
	Inoperative defrost timer.	Replace timer.
	Open defrost heaters.	Replace heaters.
Water in bottom of cabinet or ice in bottom of freezer.	Clogged drain.	Clear drain.
	Misaligned drain fitting.	Reposition fitting.
	Cabinet not level.	Level cabinet.
	Defrost timer inaccurate.	Replace timer.
	Drain heater defective or not making good contact.	Retape heater in position or replace if open. Make certain it has proper voltage before replacement.
Odor.	Food.	cover foods.
	Fouled drain system.	Clear drain system and flush with baking soda and water solution.
	Inefficient filter if used.	Check for correct installation. Change if over 1 year old.

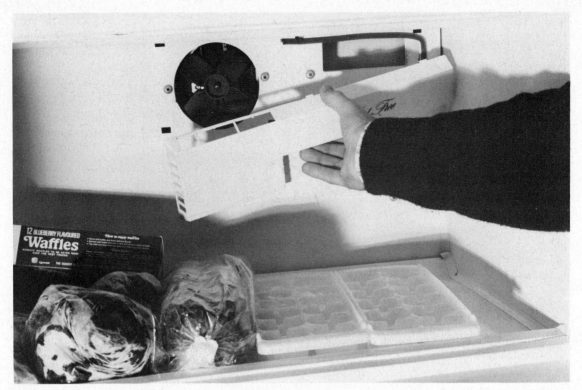

Make sure you do not block the fan in a freezer unit because that reduces its efficiency.

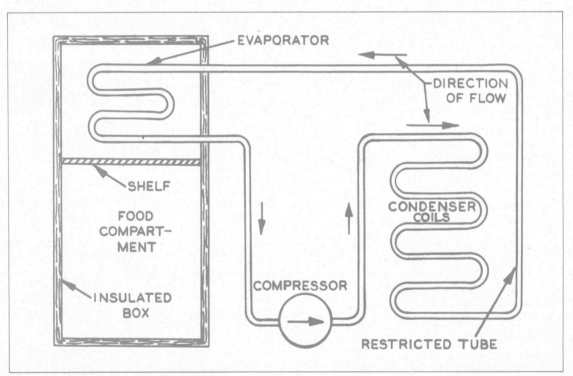

Fig. 4-4. Basic circuit of refrigerator.

that the evaporator is inside the insulated box, but all the other components are outside, some at the bottom and others at the back of the case. Following the condenser, the pipe leading to the evaporator in the insulated compartment has a restricted diameter so that the refrigerant between the compressor and the restricted tubing is under pressure. This increase in pressure condenses the vapor so that it becomes liquid again and in doing so gives off heat. The hot air rises from the coils of the condenser because it is lighter than the surrounding air, and thus cooler air moves in towards the condenser to remove more heat. In some refrigerators, this air movement is increased by the addition of a fan blowing across the coils of the condenser. From the condenser the liquid enters the restricted tubing and travels, under pressure from the compressor, to the evaporator. Here the coil diameter is increased so that the liquid entering the evaporator does so at reduced pressure. The pressure in the evaporator is further decreased by the compressor sucking out the vapor at the other end. The low pressure causes the refrigerant to evaporate quickly, drawing heat from the food stored in the refrigerator.

The temperature inside the refrigerator is controlled by a thermostat which is different from the bimetallic units used in most electrical appliances. The refrigerator thermostat has a bulb sealed to a bellows and containing a small amount of refrigerant. When the temperature increases, the refrigerant boils and causes the bellows to expand, closing a switch controlling the compressor motor. When the compressor has been running, the refrigerant in the sealed chamber contracts, and the bellows also contracts, causing the switch to open and turning off the compressor. Provision is made for the switch to open and close with a snap action to prevent arcing at the contacts.

When the compressor is off, the pressure in the tubing equalizes quickly. Thus, when the compressor motor is started, there is very little load, and a motor with low starting torque can be used. The most common type is a split-phase motor. The motor and compressor are usually sealed in a single case with

enough oil to run for ten years or more. If trouble develops in this unit, the whole thing could be replaced, but the home handyman should not try to disassemble the unit. In general, by the time motor troubles develop, the refrigerator will have given many years of satisfactory service, and purchase of a new, improved model usually will prove to be more economical than trying to replace the motor.

Because of their size and weight, refrigerators do not lend themselves to easy servicing. Nevertheless, there are some things you can do, and you should not give up as soon as trouble appears. Fortunately, refrigerators last for a long time and when they do develop big troubles, the home owner is ready to buy a new one. The repair of small problems is within your capability.

If the refrigerator fails to operate, and the light inside does not light, the trouble is almost certainly in the receptacle or line cord. Check the receptacle for voltage. Check the line cord for continuity. Sometimes a loose line cord connector falls out of the receptacle.

Refrigerator works, but interior light stays off. There are two possibilities here: either the bulb is burnt out or the door switch is broken. Replace a burnt-out bulb, as required. If the switch is bad, replace it if you can get at it

Refrigerant and container.

easily. On some models, however, replacing the switch requires much physical labor taking apart the box. You may decide to live with the bad switch, since the refrigerator works otherwise.

If the interior light works, but the refrigerator doesn't operate, the trouble may be due to a defective thermostat, bad contacts on the motor switch or broken wires. Make a visual check for broken wires or thermostat troubles. Replace these parts as necessary. Some refrigerators use a capacitor-start motor, and the capacitor is easily accessible. Try spinning the motor with your hand while it is plugged in. If it then runs, the trouble could be the capacitor, a bad starting coil or an open centrifugal switch. You can't do anything about the coil or other motor troubles, but you can replace the capacitor or switch.

If the refrigerator runs too long and the cabinet is warm, the door gaskets may be dried or broken. Put a piece of paper between the door and the frame and close the door. A dollar bill is the right size· Now pull on the piece of paper. There should be some resistance. If not, replace the door gasket. Another possiblity for this trouble is poor insulation after many years of service. This is not worth fixing. The interior can also be warmed if the light does not go out when the door is closed. Check this by pushing the door switch. If the light stays on, replace the switch, if possible, or remove the bulb.

If the refrigerator runs too long and the cabinet is too cold, the problem is a defective thermostat, or switch contacts welded closed. Either part can be replaced.

If the refrigerator is noisy, look for the source of the noise. Loose condenser tubing may be vibrating against the frame. If the unit uses a fan to cool the condenser, check the fan blades for freedom and the fan bearings for wear. If the compressor is noisy, a possible source of the trouble is too much dirt around the condenser. Clean this area with a vacuum cleaner.

The operation of the refrigerator is affected by the weather. In hot weather, the unit will work harder to pull the temperature down, especially if the door is opened frequently. Operation is also affected by the amount of food inside. With a large quantity of food, a small amount of heat removed from each item can be sufficient to heat the refrigerant, and thus the refrigerator may not get cold enough. With an almost empty refrigerator, the required heat must come from fewer foods, and consequently the small amount of food may even be frozen.

Freezers and combination refrigerator-freezers work on much the same principles. The troubles you can fix are fixed in the same manner as those in refrigerators. Freezers should last indefinitely with no major troubles.

4-3. Dishwashers

In an automatic electric dishwasher, hot water and detergent are sprayed over dirty dishes with sufficient force to dislodge food particles. Dishes are then rinsed in clear, hot water. The process is repeated, and after several clear rinses, a Calrod heating element is turned on to warm the interior and dry the dishes. Once the dishwasher is started, each process in the cycle is controlled automatically by a timer of the sort discussed in Chapter 2 and illustrated in Figure 2-7.

Dishwashers may be built into the wall cabinets in the kitchen or may be portable, mounted on wheels, to be moved near the sink when needed and then wheeled out of the way. Some models, especially portables, are loaded from the top, but built-ins usually have a hinged door on the side. When a side door is used, there is always a door interlock, or switch, which shuts off the dishwasher whenever the door is opened.

The agitation of the water in the dishwasher is provided in one of two ways, either by an impeller or a spray arm. These two types of devices are illustrated in Figures 4-5 and 4-6. Outside of the method of agitation the two types are similar. The motors are usually split-phase, although some models use shaded-pole motors.

In both types, the first operation in the washing process is draining the tub in case water was left from the last washing. All steps

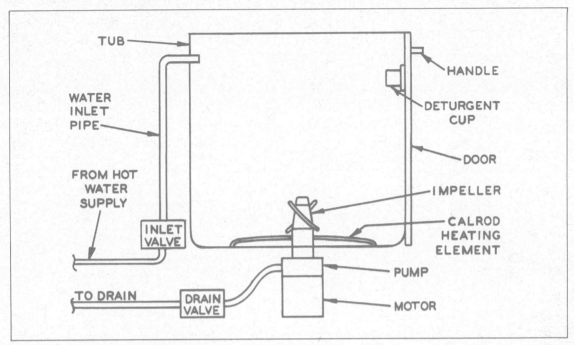

Fig. 4-5. Impeller-type dishwasher.

are controlled by the timer. The drain valve is opened, and the drain pump sucks water out of the washer and pushes it down the drain. In

Figure 4-5, a single motor and pump are shown, which is typical of impeller-type washers. In Figure 4-6, two motors are shown,

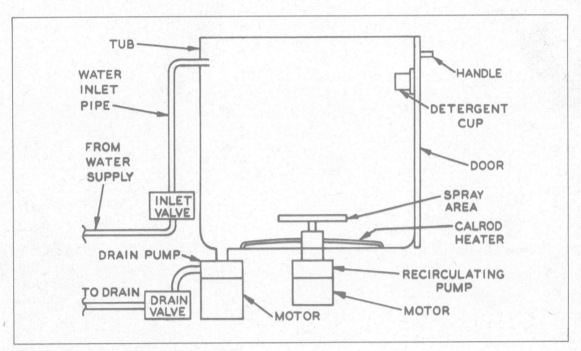

Fig. 4-6. Spray-arm dishwasher.

Interior of dishwasher.

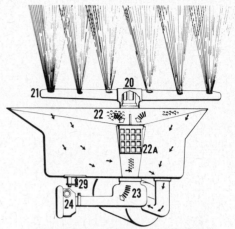

Water action for the spray type dishwasher is from the pump up through the hub (20) out the swirl arm (21) which rotates on the hub. The return path is through the filter screen (22) and scrap basket (22A) to the pump. During the drain cycle the water travels through the drain boot (23) to the drain pump (24) and out.

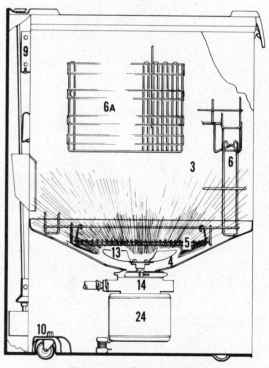

one for running the drain pump and one for running the circulating pump. The two pumps are needed on spray-arm dishwashers and two separate motors are usually used, although one motor and a system of gears and clutches could be used. Draining lasts less than a minute.

After the drain operation, the drain valve is closed, and the water inlet valve opens. This latter valve is in the hot water line. Water should be between 140 and 160 for proper performance. Water enters near the top of the tub, although the inlet valve is usually located under the tub near the motor. During the fill operation, the motor or motors are usually at rest. The inlet valve remains open for a fixed time interval set by the timer. The amount of water that enters is a function of the water pressure, and if the pressure were too great, too much water would enter and possibly leak

Water action for the splash type dishwasher is from the impeller (13) through the impeller guard (5) and against the dishes in racks (6 and 6A), then by gravity back to the pool from which the impeller picks up the water. The water level should not be more than $1\frac{1}{2}$ inches above the bottom of the impeller when the machine is in use.

Always scrape off excess food from dishes before placing them in the dishwasher.

Troubleshooting Chart   DISHWASHERS

PROBLEM	POSSIBLE CAUSE	CORRECTIVE ACTION
Machine will not run at all.	No power to machine. Loose leads. Door switch not operating. Manual reset open. Timer not operating.	Check for power at outlet. Check and secure all leads. Check door adjustment and continuity through switch. Push reset button. Check out timer.
Does not make a complete cycle.	No power to component. Timer working erratically. Loose lead. Defective component.	Check for power at component. Check operation against sequence. chart on wiring diagram. Check and secure all leads. Check for proper operation.
Water does not enter machine.	Supply valves closed. Open circuit in wiring. Timer not operating. Solenoid not operating. Supply line restricted.	Open valves. Check continuity. Check out timer. Check leads and operation. Check for kinks and foreign matter in lines.
Water does not drain from machine.	Restricted lines. Pump jammed. Motor not reversing (some impeller machines).	Check for kinks and foreign matter. Remove foreign material. Check timer switching (pump should run counterclockwise for pump-out).
Water leakage.	Poor door seal. Splash at fill valve. Split hose or loose clamps. Overfill (undercounter machines). Tub leaks.	Adjust gasket, latch. Check level of machine. Check alignment and tube end for burrs or deposits left by hard water. Check condition of hose and clamps. Check operation of pressure switch timer and inlet valve. Repair with patch kit.
Water leakage in gravity-drain.	Timer switch sticks in closed position. Leak in tub. Drain plug leaks. Drain plug sticks open. Solenoid plunger sticking because of dirt or bent bracket.	Replace timer. Repair or replace tub. Check "O" ring for damage or foreign particles. Clean or replace "O" ring. Clean lime or foreign matter from plug. Also, clean inside of drain opening. Remove plunger and coil and clean, straighten, or replace bracket.

Troubleshooting Chart   DISHWASHERS (continued)

PROBLEM	POSSIBLE CAUSE	CORRECTIVE ACTION
Unsatisfactory drying.	Low water temperature.	Should be 140° in machine at last rinse.
	Heater element inoperative.	Check electrical circuit; replace if defective.
	Impatient user.	Wait for end of cycle.
Poor washability.	Improper water level.	Check flow, washer water pressure, and installation of drain.
	Improper water temperature.	Check temperature in machine during last rinse (140-160°)
	Undesirable water conditions (hardness or excess iron).	See about a water softener.
	Improper leading.	Use proper procedure.
	Pump motor not operating (defective starting relay).	Check for restrictions; replace electrical circuit if necessary.
	Spray arm not turning.	Clean under bearing, check clearance between arm and basket rail, etc.
	Impeller loose or damaged.	Tighten and/or replace.
	Detergent dispenser not dumping.	Check solenoid; adjust linkage; avoid blocking cups.
	Loose or dirty filter screen.	Refit screen to eliminate gaps; wash out well.
	Drain pump inefficient.	See if water is being properly evacuated. Look for kinks, obstruction, foreign matter.
	Back siphoning from sink.	Check installation drain loop and/or air gap.
	Incorrect timer function.	Check timer against sequence chart.
Abnormal noise (other than water hitting dishes or small item knocked loose).	Spray arm or impeller hitting.	Check clearances.
	Foreign matter in tub.	Check for broken dish.
	Loose parts.	Check impeller and pump operation.
	Low water level.	Check for proper fill and water pressure.
	Improper loading.	Read owner's manual.
	Machine not level and solid.	Level and fasten.

around the edge of the door. To prevent this, many models also have a float valve which is in series with the inlet valve and closes when the water reaches its maximum permissible level. No more water can enter even though the inlet valve stays open for the whole time interval set by the timer.

After the inlet valve closes, there should be about three quarts of hot water in the tub. Here the two types of washers differ slightly in operation. In the impeller type, shown in Figure 4-5, the water level is above the top of the impeller. The motor is switched on and turns the impeller, which splashes water over and around the dishes with great force. Some of the water mixes with the detergent contained in an open detergent cup in the door of the washer. As this water falls back and is splashed out again, all the water gets mixed with detergent. After this wash operation, which lasts five or six minutes, the impeller stops, and the motor drives the pump to drain out the water through the drain valve, which is now open.

This modern dishwasher is an eight-cycle model.

In the spray-arm dishwasher, the water level is below the spray arm. The recirculating pump drives the water up through the spray arm, and out openings in the arm. Because these openings are offset, the arm revolves and sprays all over the interior and mixes water with detergent just as in the impeller type. At the end of five or six minutes the recirculating pump is shut off, and the drain pump starts and drains out the water.

Clean water enters and a rinsing operation takes place, since the detergent cup is now empty. This lasts three to five minutes.

After the rinse water is drained out, a second detergent cup opens, and another wash operation begins. This may last up to fifteen or sixteen minutes. The dirty water is removed, and two or more rinses with clear water follow. Finally, all the water is drained out, the motor is turned off, and the Calrod heater turns on to dry the dishes. If you open the door of the dishwasher soon after the end of the cycle, *do not touch* the Calrod heater, since it will still be hot.

A common complaint is that dishes are not cleaned thoroughly in an electric dishwasher, but more often than not this is caused by improper use. First you must consider the nature of the "dirt" on dirty dishes. Those foods that are soluble in water, such as sugar, syrups, fruit juices, and raw eggs are easily washed in the dishwasher. Also, oils and fats are melted by the hot water and emulsified with the detergent, so that these present no problems, either. Foods such as bread crumbs and gravies are not soluble, but the scrubbing action of the spray usually breaks up the particles and removes them from the dishes. Some foods, however, *if allowed to dry on the dishes*, are very difficult to remove either by hand or in a dishwasher. These include cooked eggs, tomato juice, apple-sauce. If these foods are served, the dishes should be soaked until they are ready to be put into the machine. Also, when placing dishes in the dishwasher, you must be sure that they face the direction of the spray, that they are not nesting together, and that the detergent cup is not blocked from the spray by a large dish.

Improper washing can also be caused by

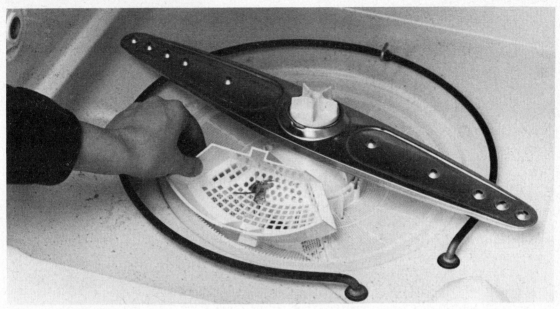

Cleaning filter on dishwasher.

water being the wrong temperature. Make sure it is in the range of 140 to 160.

If the washer takes too little water, it can also cause unsatisfactory cleaning. If too much water enters, it may leak out around the door. In either case the trouble may be in the timer or the intake valve. Check the valve solenoid for continuity. Check that the valve moves easily. Timer troubles and their corrections are discussed in the chapter on control devices.

If the drain valve leaks, it may be due to a particle of food holding the valve open or to a defective solenoid. Move the valve lever open and shut several times to clear the particle. Check the solenoid for continuity. A leaky inlet valve is caused by a bad solenoid or a defect in the valve itself. Both drain valves and intake valves are easily replaced if defective. Solenoids can be replaced without replacing the valve.

If the washer continues to operate when the door is opened, the interlock switch is defective. Clean the contacts, if possible, or replace the switch.

Noisy operation may be due to a broken impeller, interference or improper stacking of the dishes. This is always a mechanical difficulty and locating the source of the noise usually pinpoints the solution.

If the washer will not start, assuming the power line is live, the trouble may be in the interlock switch or the timer. First, check the switch and the timer motor for continuity. Finally, make sure the timer is not jammed, and replace defective parts.

If the main motor will not run, it may have a burnt-out winding. In the case of impeller-type machines, if the impeller is stuck because of improper loading of dishes, the motor will not turn. Make sure the impeller is free. Check the continuity at motor terminals. Motors should last for many years, but eventually a bearing leaks and water gets in, ruining the motor. Replacing the motor is a big job because of its inaccessibility. It is necessary to pull out the dishwasher, after making sure it is unplugged, and tip it on its side to reach the motor. With help, it can be done.

If the dishes do not dry, the Calrod heater is not working. The heater itself may be defective or it may not be getting voltage. Check the heater for continuity. If it is open, replace it. If it is good, check the timer switch which turns on the heater. Also look for loose connections.

Make sure the cabinet is grounded to prevent electric shocks. It should be grounded when it is first installed, but if you pull it out for a major repair, you might break the ground wire. *Always check for safety after servicing.*

4-4. Clothes Washers

The process of washing clothes involves many separate and distinct operations. First, the clothes should be soaked to loosen as much dirt as possible. The dirty soak water should be removed and fresh water added. Some sort of beating action now takes place, similar to scrubbing clothes on a washboard, to make sure the soap or detergent penetrates to all parts of the clothing. The water is removed, and this is followed by one or more rinses with clear water. After the last rinse, the water is removed, and the clothes are dried.

In an automatic clothes washer, all these operations are performed in the proper order and at the proper time. The clothes do not come out completely dry, but the final operation is a rapid spin which removes most of the water by centrifugal force. Manufacturers have designed many different types of washers to perform these operations, but they fall into two categories: (1) washers with cylindrical drums, and (2) washers with agitators.

In the cylinder type of washer, the clothes

Clothes washer and dryer.

are placed in a cylindrical drum which rotates about a horizontal axis. The cabinet in which the cylinder is mounted is essentially a fixed tub, and the cylinder is perforated, so that when the tub is full of water, the water will surround the clothes inside the cylinder. A motor beneath the tub is coupled to the cylinder through a belt drive or set of gears. The access door to the cylinder is usually at the front of the cabinet, and the clothes are placed into the end of the cylinder, as depicted in Figure 4-7 (a). This is called an end-loaded or front-loaded washer. In some cylinder-type machines, top loading is used: the cylinder is closed at the ends but has a door on its circumference, as shown in Figure 4-7 (b).

The agitator-type washer is shown in Figure 4-8. A perforated basket mounted on a vertical axis replaces the cylindrical drum, and an agitator is added at the center. The motor, at the proper times, is coupled to the basket, to the agitator, or to the drain pump, as required.

In both types, a timer controls the operation. This is a clock motor with cams on its shaft as described in the chapter entitled "Control Devices". First, the inlet valves are opened so that hot or cold water or the proper mixture can enter. After a predetermined time interval, the inlet valves are closed, and the clothes are "scrubbed". In the agitator-type machine, the agitator moves back and forth about its axis so that the clothes are tossed about in the water. In the cylinder-type, the scrubbing action is caused by baffles on the

(a) END LOADING

(b) TOP LOADING

Fig. 4-7. Cylinder-type clothes washer.

Time and water usage chart. (Courtesy *Norge*)

REGULAR CYCLE	TIME	GALLONS	GENTLE CYCLE	TIME	GALLONS
Fill	4	11.2	Fill	4	11.2
Wash	2-10		Wash	3	
Pause	1		Pause	1	
Spin	2		Spin	2	
Spin Spray	1	2.8	Spin Spray	1	2.8
Fill	3	8.4	Fill	3	8.4
Overflow Rinse	4	11.2	Overflow Rinse	1	2.8
Deep Wave Rinse	1		Deep Wave Rinse	1	
Pause	1		Pause	1	
Spin	1		Spin	3	
Spin Spray	1	2.8			
Spin	5				
TOTAL	34 min.	36.4 gal.	TOTAL	20 min.	25.2 gal.

inside of the cylinder carrying the clothes around with them until they fall to the bottom as they near the top of the arc. The soaking and agitation may take ten or twelve minutes. Then the water is spun out by rotating the basket or cylinder at high speed, and the pump empties the water out of the tub. The drain valve is opened for this operation and closes again when it is over. This is followed by more fill-agitate-spin operations. In some machines extra rinses are provided. Finally, the clothes are spun, and everything is turned off. The intake and drain valves are solenoid operated, as described in the chapter on control devices. The solenoids are controlled by the timer, which also controls movements of gears and belts, so that the motor drives the proper pump, agitator or basket.

When something goes wrong with a washing machine, even if you diagnose the trouble, you may wonder how to take the panels and top off so that you can make the necessary repairs. Disassembly techniques vary from machine to machine. If there are no screws in view holding down the top, slide the top toward the front and it can be removed. The sides and front and back can then be removed by taking out screws holding them. On agitator machines, you must remove the agitator to get at the tub. The top of the agitator can be unscrewed from the shaft, and the agitator lifted out. If it sticks, loosen it by tapping on the shaft lightly with a rubber mallet while exerting upward pressure on the agitator. After the agitator is out, the shaft can be removed by loosening some Allen head

screws near the base. After this, all disassembly is straightforward. However, do keep a record of how things are taken apart so that you can reassemble them correctly. Important: when lifting off the part containing the timer, handle it carefully.

A common trouble in a washing machine is a jamming or blockage caused by a piece of clothing falling out of the basket or cylinder and getting stuck. It can jam the works, stopping the machine, or can even deflect the water so that it spills on the floor. Always suspect a sock or handkerchief whenever the machine develops such trouble.

In most machines, hoses from the hot and cold taps are permanently connected, and

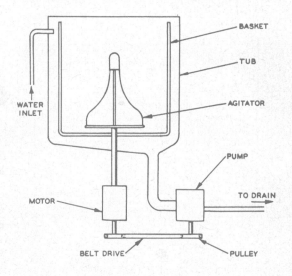

Fig. 4-8. Agitator-type clothes washer.

Troubleshooting Chart   AUTOMATIC WASHERS

PROBLEMS	POSSIBLE CAUSE	CORRECTIVE ACTION
No water fill.	Water valves closed. Hoses kinked. Screen in fill hose clogged. No power to fill solenoid. Faulty water level control. Machine not turned on. Machine did not drain out last time used.	Turn on valves. Unkink hoses. Clean out screen. Replace solenoid. Replace control. Check controls and power at outlet. See "Water will not drain from machine."
Incorrect fill or temperature.	Faulty water level control. Faulty thermal element in mixing valve. Hot-water supply inadequate. Reversed hoses—hot-water hose on cold-water connection.	Replace control. Replace valve (sometimes repair kits are available). Check temperature setting and capacity. Connect hoses correctly.
No spray rinse.	No water supply. Defective timer.	Same as no water fill. Replace timer.
Water will not shut off.	Defective timer. (Time fill machines) Defective water level control. Foreign particles in mix and fill valve. Defective valve.	Replace timer. Replace control. Clean out valve. Replace valve.
Water leakage.	Inlet hose loosely connected to valve. Drain hoses not tight on pump. Broken hose. Leaky gasket. Cracked housing.	Tighten hose connection. Tighten hose clamps. Repair hose. Replace gasket. Replace parts.
Water will not drain from machine.	Kinked or clogged drain hose. Pump does not run. Suds lock. Faulty transfer valve. Defective timer. Loose belt.	Clear drain hose. Readjust and tighten pump. Drive mechanism. Remove suds, add cold water. Replace valve. Replace timer. Adjust belt.
Motor will not run.	No power to machine. Door switch or other safety control in motor circuit. Faulty timer. Faulty water level control. Faulty motor.	Check outlet. Check controls for operation and replace if defective. Replace timer. Replace control. Repair or replace motor.

Troubleshooting Chart   AUTOMATIC WASHERS (continued)

PROBLEM	POSSIBLE CAUSE	CORRECTIVE ACTION
No agitation.	Motor failure. Faulty timer contacts. Faulty transmission. Defective control solenoid. Broken linkage. Faulty water level switch.	Repair or replace motor. Replace timer. Repair or replace. Replace solenoids. Replace or repair linkage. Replace switch.
Slow spin.	Belt or clutch slips.	Adjust belt.
Excessive vibration.	Washer not level. Weak flooring. Unbalanced load. Rubber cups not on feet. Damaged snubber or suspension bolts.	Level machine by adjusting leg screws. Reinforce floor. Redistribute load. Install cups on feet. Replace snubber.
Torn clothing.	Improper bleach usage. Broken agitator. Defective basket.	Add bleach to water before loading clothes in tub or dilute bleach well before adding. Replace agitator. Replace basket.
Machine will not shut off.	Defective timer. Break in wiring.	Replace timer. Repair wiring.
Timer will not advance to next cycle.	Defective timer motor. Bound timer shaft or knob. Faulty water level control.	Replace timer motor. Clear knob from panel. Replace control.
No suds return (for machines with suds saver).	Faulty distributor valve. Slipping belt. Kinked hose. Defective solenoid. Faulty water switch. Jammed pump.	Replace valve or solenoid. Tighten or replace belt. Straighten hose. Replace solenoid. Replace switch. Remove pump and clean out.
No recirculation of water during agitation.	Jammed pump. Defective pump drive. Clogged hose. Defective distribution valve.	Clean out pump. Replace coupling or tighten. Clean out hose. Clean out or replace valve or solenoid.

Note: Timer does not advance during water fill period until the water level switch has been satisfied.

Lint trap on clothes dryer.

these taps are left open. When the intake valves open, water flows through these hoses, but when the valves are closed, the water in the hoses is under pressure from the water mains. Presumably the hoses could rupture under this pressure, although it rarely happens. However, to prevent possible ruptures, especially when going on vacation, many users shut off the faucets that feed the washer, and then they forget to turn them back on and wonder why no water enters the machine. If no water enters, check the faucets. If the faucets are open, the trouble may be due to a bad solenoid on the inlet valve, a defective valve, or a fault in the timer. Tests for these faults and corrections are described in the chapter on control devices.

If the motor does not run, make sure voltage is getting to the motor. Check the motor for continuity. The motor is usually a capacitor-start type, although split-phase motors are also used. Remove the load from the motor and check it by the techniques described in the chapter on electric motors. If necessary, it can be replaced. However, washing machine motors usually outlast most other parts. If the motor finally fails, other parts of the washer may be on their last legs, and it may not be worth replacing the motor.

If the washer fails to perform one of the required operations in the washing cycle, check the timer, relays and wiring. Look for loose or broken belts, or grease on clutches. Most machines have a switch which is actu-ated by the lid so that agitation stops when the cover is opened. If this switch is defective, the washer will not agitate. You can check the switch by a continuity test across it, with the washer unplugged.

If the washer is noisy or vibrates excessively, the clothes are probably distributed unevenly in the basket or drum. Redistribute the clothes.

If the washer leaks, check all gaskets, seals and hose connections. Also check hoses for breaks. Replace or tighten as required.

It is most important that the washing machine be grounded, since it is used next to a good ground, the cold water pipe. After you have finished servicing it, make sure the ground wire is still connected. Also, as a precaution, measure the voltage between the case and ground both while the machine is running and when it is off. *The voltage must be zero.*

4-5. Clothes Dryers

In a clothes dryer, damp clothes taken from a washing machine are tossed about in a current of heated air until they are completely dry. Although dryers take many forms, all operate on much the same principles. The clothes are placed in a perforated drum with baffles, as shown in Figure 4-9. As this drum rotates, the clothes are tumbled about.

Motor of Automatic dishwasher.

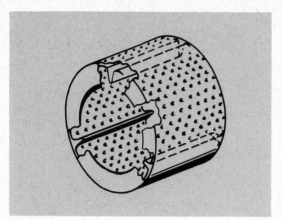

Fig. 4-9. Tumbling drum.

The components of a clothes dryer are shown in Figure 4-10. The drive motor is a split-phase type of about 1/3 horsepower, designed to run on 115 volts, AC. A large pulley on the drum is coupled to a pulley on the motor by a belt so that the drum rotates at about 50 rpm, much slower than the motor. Another belt couples the motor to a fan which sucks in air from outside and blows it over a heater and through the drum. The damp air is then blown out through a vent usually leading outside the building.

The source of heat in a dryer may be either gas or electricity. In a gas dryer, there must be some means of igniting the gas when it is turned on, and this is usually done by electricity, although in some models a pilot burner is on all the time. The main burner is located in the path of the air being blown toward the clothes. In an electric dryer, the heating element is usually located in a strip around the entire perimeter surrounding the drum. The heater is a coil of nichrome wire which operates on 230 volts and puts out about 5000 watts. Thus, an electric dryer must be connected to a 230-volt line. The motor, however, runs on 115 volts, which is obtained between one side of the 230-volt line and ground. A gas dryer, of course, must be connected to a 115-volt line to run the motor.

In a gas dryer, if the flame is blown out accidentally, it is important to close the gas supply, since gas fumes are dangerous. A special circuit is used to prevent gas flow when there is no flame, as shown in Figure

4-11. When the dryer is turned on, current flows through a platinum heating coil (not shown in the figure) located just above the pilot jet. At the same time this current flows through the solenoid which opens the gas line to the pilot jet. Gas flows out of the jet and is ignited by the heating coil. A *thermocouple* is located above the pilot flame. A thermocouple is a device which produces a voltage when heated. The voltage produced in the thermocouple by the pilot flame actuates the solenoid valves to hold open the gas lines to the pilot and the main burner. The current to the heating coil is shut off, also removing the original current to the solenoid. The solenoid valve, however, is now held open by the thermocouple. If the flames blow out, the thermocouple would cool and no longer provide a voltage to actuate the solenoids. The solenoid valves would close, shutting off the supply of gas. The voltage developed in the thermocouple is much less than one volt, but is sufficient to hold the solenoid valves open.

When a dryer is turned on, the drum should start revolving immediately. Heat is not applied, however, until the drum is almost up to full speed. This is a safety precaution, since

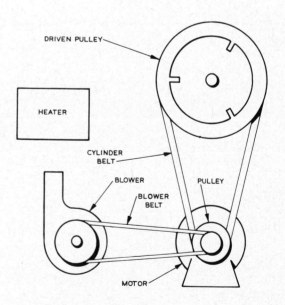

Fig. 4-10. Dryer components.

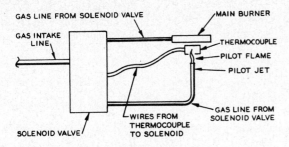

Fig. 4-11. Gas controls.

if the clothes did not tumble, the heat could be concentrated on one article of clothing, perhaps burning it. A centrifugal switch is used to start the heat *after* the motor is rotating fast enough. The basic circuit in an electric dryer is shown in Figure 4-12. A centrifugal switch is in series with the heater across the 230-volt line. This switch is mounted on the motor and is tied to the switch which opens the starting coil circuit. Thus, when the motor gets up to speed, centrifugal force opens the starting coil switch and closes the heater switch simultaneously. In a gas dryer, a similar centrifugal switch is used in series with the main gas solenoid. Thus, the solenoid valve does not open to permit gas flow to the main burner until the motor is up to speed. Note that it is possible for the motor to be running while the drum is stationary. This could happen, for example, if a belt broke. In this case, the heat would be turned on as soon as the motor got up to speed even though the clothes would not be tumbling. For this reason, the user should always make sure the clothes are tumbling when the machine is first turned on.

Some of the operations in a dryer are controlled by a timer similar to that described in the chapter entitled "Control Devices". The user turns a knob setting the length of time of operation, and as the dial turns back to zero, cams operate the proper circuits. The motor is turned on immediately, turning the drum and the blower fan. These remain in operation throughout the cycle. A switch actuated by the timer turns on the platinum heating element which ignites the gas in a gas dryer. After a suitable interval, the switch is opened

again by the cam in the timer. About ten minutes before the end of the drying cycle, the heat is shut off by the timer. This allows the clothes to cool somewhat so that they can be handled when the dryer stops.

Dryers are equipped with safety devices to prevent damage to clothes, to the machine itself, and harm to the user. The door has a switch which is in series with the main line, and shuts off the machine when the door is opened. In addition, thermostats prevent the interior from getting too hot by turning off the dryer when the air reaches an abnormally high temperature.

Most common faults in a dryer are mechanical. If the dryer motor runs, but the drum doesn't turn, the trouble could be a loose or broken belt or a pulley slipping on its shaft. Make sure the drum is free to turn. Sometimes an article of clothing can jam it. Then check belts and setscrews on the pulleys. Replace broken belts as required.

If the dryer operates, but the clothes remain damp, either there is no heat or the fan is not blowing air through the clothes. Check the fan belt first. Also check the fan bearings to make sure the blower is free to turn. If the air coming through is cold, the trouble could be in the heat source or switches. If the heater in an electric dryer is defective, it cannot be repaired, but you can replace it with a new

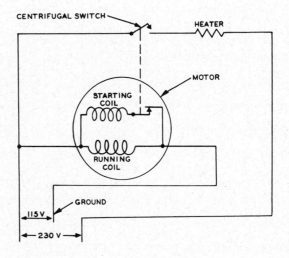

Fig. 4-12. Heater delay circuit.

Interior of garbage disposer.

Baseboard heater.

Troubleshooting Chart   ELECTRIC DRYERS

PROBLEM	POSSIBLE CAUSE	CORRECTIVE ACTION
Will not run.	No power.	Check fuse and power supply.
	Loose wiring.	Check terminals and wiring.
	Door switch.	Make certain door closed properly to actuate switch.
	Defective motor.	Check motor.
	Defective timer.	Replace timer.
Runs but will not heat.	Loose wiring.	Check terminals and wiring.
	Defective thermostat.	Replace thermostat.
	Defective centrifugal switch in motor.	Replace switch (check linkage to motor).
	Defective timer.	Replace timer.
	Open heater element.	Replace heater element.
	Heat switch set to OFF.	Set switch for desired heat.
Drum will not rotate.	Broken or slipping belt.	Replace belt.
	Jammed.	Check for foreign article between drum and shroud.
		Replace defective bearing or bearing support which allows drum to sag and hit.
	Loose pulley.	Tighten pulley set screw.
Clothes not drying, but dryer runs.	Defective operating thermostat.	Replace thermostat.
	Fan loose on shaft; no air motion.	Tighten set screw.
	Clogged lint screen.	Clean out.
	Leaky door seal; air leaks.	Replace door seal.
	Incorrect heat or timer selection.	Reset timer and/or heat control.
	Clothes too wet when placed in machine.	Wring out or extract water before placing in dryer.
Will not shut off.	Defective timer.	Replace timer.
Blows fuses	Electrical ground.	Check heater element for foreign matter or sagging drum.
		Check wiring for bare spots touching the frame.
Motor runs when door is open.	Defective door switch.	Replace door switch.
Bulbs do not light.	Defective bulb.	Replace bulb.
	Loose wiring.	Check and reconnect.
Timer fails to advance.	Dial binds.	Relocate on shaft.
	Timer motor defective.	Replace timer motor.
	Door switch open (same models).	Close door or replace door switch.

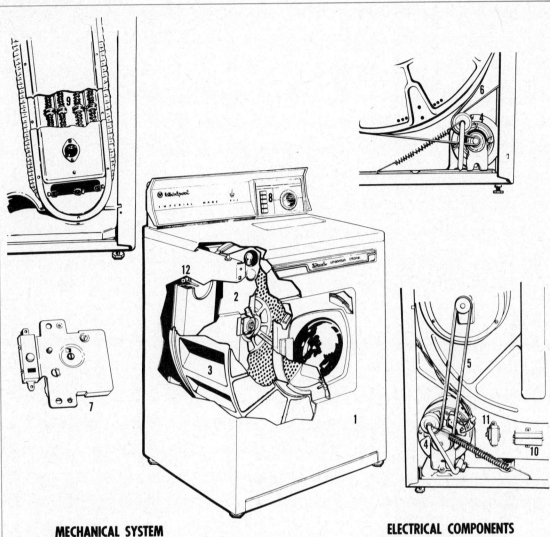

MECHANICAL SYSTEM

- **CABINET (1)**
  U-Type
  Removable front panel
  Hinged Top
  Glass port in door
  Porcelain bearing blocks

- **BULKHEAD (2)**
  Removable - Drum shaft welded to bulkhead

- **DRUM ASSEMBLY (3)**
  Rolled seam
  Stamped baffle
  Front and rear bearing support
  Rear self-aligning bearing
  Delrin bearing ring - front

- **DRIVE SYSTEM**
  Double end Shaft Motor (4)
      Two poly "V" belts
  Direct Blower Drive (5)
  Direct Drum Drive (6)

ELECTRICAL COMPONENTS

- **TIMER ASSEMBLY (7)**
  Automatic Master Touch Control Button
      Closes start switch to machine motor
      Depress timer reset button
      Ten-Minute timer motor operation in a
      cycle

- **PUSH BUTTON SWITCH (8)**
  Has five buttons - Controls four cycles and
  console light

- **HEAT ELEMENT (9)**
  Two Sections - One 2,000-watt and one
  3,600-watt - three lead terminals

- **HOT WIRE RELAY (10) and
  TRANSFORMER (11)**
  Relay has an 11-volt hot wire
  element--
      Operates contacts from contraction and
      expansion of hot wire.

  Transformer--
      Step down 230 to 11-volt output

- **SENSOR THERMOSTAT (12)**
  One inlet - heater box - 200° reset, normally
  closed
  One exhaust - sensor fan scroll - 145° open,
  normally closed

Mechanical system and electrical components of a typical clothes washer.

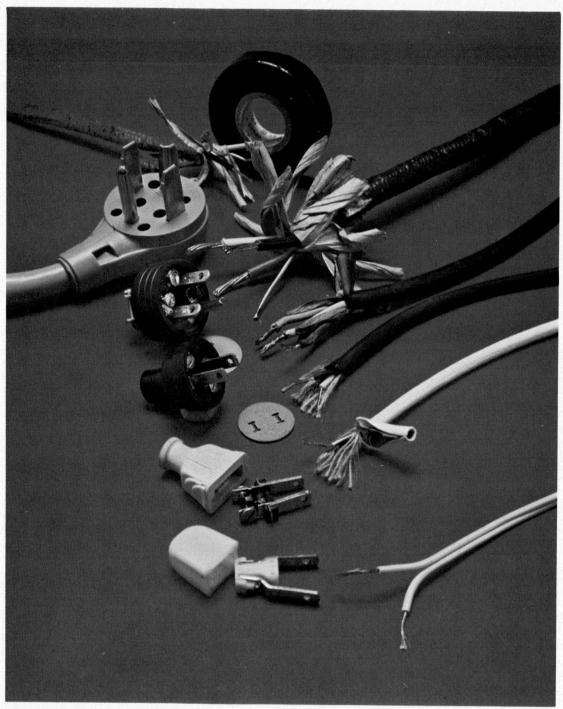

Types of line cords and plugs.

one. In a gas dryer, the most frequent trouble in the gas line is failure of the solenoid valve which actuates the pilot. Another trouble is a burnt-out heating element for the pilot. In either case, replace with an identical part.

If the dryer won't run at all, first make sure the outlet is live. Line cords rarely give trouble, since they are never disconnected or moved. If there is voltage at the outlet, check the door switch, the timer, and finally, the motor. Check all connections. Make continuity and voltage checks to ascertain if voltage is getting to the motor and timer. For motor troubles and their cures, consult the chapter entitled "Electric Motors".

If the clothes dry too slowly, they may have been too wet when they were put in. This difficulty can also occur if the dryer is overloaded. Also, check for restriction of the air path. If the user fails to clean the lint trap, it may eventually get clogged and block the flow of air. A loose belt can also reduce the flow of air.

If the trouble is in the gas line, the trouble may be a solenoid valve, as mentioned earlier, or it may be a defective thermocouple. You can call the gas company, and in most areas they will send a representative to determine the source of trouble. However, he will not fix it. If he identifies the fault as a defective thermocouple, you can replace it yourself.

A thermocouple looks like a long copper tube with a small bulb at one end. When heat is applied to the bulb, a voltage is produced at the other end. Usually the thermocouple can be removed by means of a crescent wrench and replaced with a new one. Thermocouples are used in gas water heaters and gas furnaces, as well as gas dryers, and can also be readily replaced in these appliances when they are defective.

4-6. Electric Ranges

One of the most heated controversies in the appliance field has been on the merits of gas versus electric ranges. Each has fine features.

Frequently the final choice by the home owner is influenced by a personal preference that is stronger than the actual difference between the two types of stoves.

Some characteristics of the electric range, such as slow heat-up and slow cool-off, are regarded as being an advantage or a disadvantage, depending on one's point of view. Similarly, the gas stove is alternately pointed to with pride and viewed with alarm. It has, however, been wisely pointed out that a meal cooked on one type is just as good as one cooked on the other.

The electric range can be regarded as a hot plate that has become sophisticated. Basically, the operating principle is simple: heating elements are switched on and off. It is the ensemble of gadgets that makes the difference — lights, timers, temperature sensors, and all the rest of the accessories that go into the modern range.

The range-heating element is the product of many years of design effort. It heats and cools quickly, is mechanically rugged, and relatively impervious to food and water spillage. It is a hermetically sealed unit that consists of a *nichrome* resistance coil imbedded in an insulating powder (usually magnesium oxide) that is contained in a stainless-steel tubing. Because of the insulating material, no part of the coil contacts the inner surface of the protective sheath. Such a contact would constitute an unintentional ground, in addition to producing a hot spot. Both the spiral surface unit and the curved D-shaped oven unit are constructed this way.

Depending on the type of range on which it is used, the spiral-shaped surface unit contains either a single heating coil or two such coils. Ranges with multiposition snap switches or push-button heat switches need two-coil surface units. The switch applies 115 or 230 volts to the two elements in series and in parallel, thereby providing various cooking heats. Ranges with the "infinite" or "cycle timer" control will have single-coil surface units. The oven unit, either bake or broil, is generally a single-coil, thermostatically controlled unit.

In the oven, a thermostat is used for temperature control. It is the fluid-expansion type of thermostat, not a bimetal type. A switch offers selection of upper or lower heating units, but the same thermostat exer-

Troubleshooting Chart   ELECTRIC RANGES

PROBLEM	POSSIBLE CAUSE	CORRECTIVE ACTION
Oven will not heat.	Selector switch is OFF.	Set selector switch.
	Blown fuse.	Check fuses.
	Inoperative oven control.	Check circuit continuity.
	Open circuit in oven element.	Check circuit continuity.
	Loose connection.	Tighten all connections.
	Timer inoperative.	Check timer setting.
Oven too hot or too cold.	Thermostat calibration.	Thermostat adjustment.
	Improper oven door fit.	Door seal and fit.
Oven will not turn off.	Inoperative selector switch.	Check selector switch.
	Inoperative timer.	Check timer setting.
Oven interior light does not light.	Loose or inoperative bulb.	Tighten or replace bulb.
	Inoperative light switch.	Light switch replacement.
	Loose connections.	Tighten.
Oven door opens under heat.	Door needs adjustment.	Door seal and fit.
	Loose or worn pin.	Replace hinge.
Oven door drops down.	Worn hinge bracket.	Replace bracket.
Timer does not operate properly.	Incorrect setting.	Refer to owner's manual. See "Timer operation."
	Loose connection.	Tighten.
	Inoperative motor.	Replace motor.
	Inoperative mechanism.	Replace timer.
Timer will not control oven.	Incorrect connection.	Check wiring diagram.
	Inoperative timer.	Replace timer.
	Selector switch not correctly set.	Set selector switch.
Oven drips water or sweats.	Oven not preheated with door open.	Check oven operation.
	Oven temperature excessive.	Check thermostat calibration.
	Door does not seal at the top.	Adjust oven door.
	Clogged oven vent.	Clean vent.
Surface unit does not heat.	Blown main fuse.	Check fuse.
	Loose connection.	Tighten.
	Inoperative switch.	Replace switch.
	Open unit.	Replace unit.
	Incorrect connection.	Check wiring diagram.
	Broken wire.	Continuity check.

cises temperature control for either "bake" or "broil." Upper and lower heating units are rated at 2000 to 3000 watts. When continuously energized at 240 volts, they can produce the highest temperature marked on the control knob. Lower temperatures are obtained by thermostatic cycling.

Oven timers are quite popular. Most ranges are equipped with some sort of device for automatically turning the oven on and off. Two types of timers are used: one is a clock-operated switch, the other an electric alarm bell. The switch timer actually turns the oven element on and off at a preset time. The bell merely notifies the user that a preset time has elapsed. It does nothing to control the operation of the range.

Electrical troubles in the range can generally be classified as heating element, controller, and wiring faults. In addition, such accessories as lights and timers can become defective. However, troubles that interfere with the basic action of the range are most likely to cause the owner to call for an appliance technician.

Although it is ruggedly constructed, the heating element can, with time or because of rough usage, become open-circuited. Then it will no longer heat. The heating element is easily checked with an ohmmeter. Normal resistance is somewhat less than 100 ohms. If open-circuited, it will measure ∞. During the resistance check, the heating element must, of course, be disconnected from the electric circuit. An open-circuited element will remain cold even with full voltage applied, but *testing the* resistance *in a live circuit means instant destruction* to the ohmmeter. If it is a two-coil surface unit, both elements should be tested.

Of all range repairs, heating element replacement is the simplest, primarily because all parts are easy to reach. As in any range repair, the first step in element replacement is to turn off the power, preferably at the fuse box.

Pull the fuses and put them in your pocket until the repair is completed. If you are carrying the fuses, someone else is less likely to energize the range while you have a fistful of range wires. If the power panel contains circuit breakers, turn off and cover them with tape to discourage anyone else from turning them on.

For ease of cleaning, the range-surface unit is designed to hinge up or lift out. Either arrangement offers easy access to the connector for surface-unit replacement. To remove the swing-up unit, the hinge hardware must be loosened.

A two-piece porcelain insulator, held together by screws or clips, surrounds the wire attachment to the surface unit. With the insulator removed, it can be seen that the wires are fastened to the heating element leads by screw-type terminal lugs. If a two-element unit is to be replaced, care should be taken to ensure exact relocation of the line wires. If necessary, prepare a simple sketch to ensure correct reattachment to the replacement unit. On a single-coil unit, with only two wires attached, no such problem exists.

After lead attachment, the insulators should be carefully reassembled around the connections and the surface unit relocated in the range *before* the fuses are restored and the surface unit energized.

Except for working in more cramped quarters, testing and replacing oven coils is no more difficult than servicing surface units. In fact, some oven coils are plug-in units that can be replaced without tools. As with the surface unit, some identification system should be used to ensure correct wire connections when a replacement is made. Using an exact replacement is preferable; however, some excellent universal-type replacement units are also available. When such a type is used, *the installation sheet supplied* with the unit *should be carefully followed to ensure a correct fit and proper attachment.*

If the heating element resistance is normal, yet the unit remains cool when switched ON, the controller is next in order as a probable source of the trouble. Some amount of range disassembly is necessary to make the switch available to test. For example, four screws must be removed to allow pulling out the push button switch assembly.

A voltage measurement is the most reliable test for a switch. When the switch is turned ON, a measurement across the switch terminals should register zero volts. When the switch is turned OFF, a measurement across the switch terminals should register

Electrical problems in an electric range can generally be classified as heating elements, controller and wiring faults.

full-line voltage, either 120 or 240 volts. A measurable voltage across a closed switch indicates a fault.

The infinite switch gives a somewhat different voltage indication. Normally, in the closed condition, the switch voltage measurement will fluctuate from zero to full-line voltage in a manner that depends on the switch setting. At a high-heat setting, the voltmeter will register zero for a longer time interval than the time it registers full-line voltage. At a low setting, the reverse is true. A faulty controller either registers zero at all times (contacts "stuck" cannot open) or registers some voltage at all times (contacts "burned" and are unable to close satisfactorily).

Switch replacement can involve the disconnection and reattachment of many wires. As mentioned before, some scheme of wire marking should be used to ensure correct relocation of each and every wire. A sketch or identifying tags are recommended.

The attaching hardware for rotary and thermostatic switches can usually be found under the knob. After the front screws are loosened, the controller is removed at the rear side of the control panel. Thermostat removal introduces a new problem, routing the fluid bulb and tubing. Removing and installing a new thermostat involves running the sensing element along a path that depends on the brand of range. The manufacturer's instruction normally gives the exact method of replacement.

Although it seldom happens, a defective range may be the result of faulty wiring, either a severed wire or a broken conductor under the insulation. So if the heating unit remains cold even though the element and switch test out satisfactorily, the wiring in the inoperative circuit should be checked. Usually arcing damage is evident as a telltale blackening of the wire. But if the faulty location is not visible, a test is necessary.

Because of the particular circuit that is inoperative, one or possibly two conductors will, through a process of elimination, turn out to be the faulty ones. To test a wire, disconnect both its ends from the circuit; then check it with an ohmmeter. A good wire should check 0 ohms; a faulty one, ∞.

While checking the wire, jiggle it, watching for an "intermittent" break that opens and closes, causing the meter needle to bounce back and forth from 0 to ∞. If the ends of the wire are too far apart for the reach of the ohmmeter leads, temporarily fasten one end to the frame of the range and test between the other end and a nearby bare-metal part of the frame.

Replace a faulty conductor with an asbestos-covered heater wire. Other types of insulation are not intended to withstand heat and are not suitable. Solder is not used, for it would melt. Instead, either the spade-tongue or quick-connect terminals are used. Because of the ready availability of connector kits, no problem should be encountered in attaching suitable end connectors. Any appliance parts supplier can usually supply the asbestos-covered wire for range rewiring.

It should never be assumed that a range must operate satisfactorily just because a part has been replaced. Even new parts have been known to be faulty. Also, an error can be made in rewiring or reassembly. Therefore any repair job should be validated by a performance check.

Fortunately, a range is easy to check. That the unit heats up satisfactorily is quite evident without the aid of instruments. However, the cooking temperatures should be tested and checked with a thermometer.

However, when repairs and/or replacements that involve either surface-heating units, oven-heating units, or any part of a thermostat have been made, a check should be made. Verify that the temperature shown on the control knob coincides, within acceptable tolerances, with the actual temperature of the unit or oven. If it does not, additional work will be necessary to discover and correct the difficulty.

4-7. Self-Cleaning Ovens

Even the most careful housewife will find that grease, starch, and sugar will spill in an oven and often bake on to the bottom or sides. Fortunately, these soils, all hydrocarbons, will decompose into water vapor and gases when heated to a temperature between 850 to 925° F.

Troubleshooting Chart   SELF-CLEANING OVENS

PROBLEM	POSSIBLE CAUSE	CORRECTIVE ACTION
No heat for cleaning.	Incorrect control setting.	Set controls for clean cycle for a minute or so to determine if all units are heating. *Note:* If they do, instruct user. If not, proceed.
	Open fuse.	Test and replace.
	Defective oven relay.	Replace relay.
	Defective thermostat.	Replace thermostat.
	Door not locked.	Check alignment or interference with locking mechanism.
	Defective selector switch.	Replace.
Incomplete cleaning.	Incorrect control setting. Set time too short.	Time must be set for at least 2 hours.
	Low-line voltages.	Check voltage under load; if low at fuse panel, notify local electric utility. If OK there and low (100 volts) at range, have qualified electrician check wiring.
	Defective element.	Check each for operation in bake-and-broil position. Check catalyst unit for continuity.
	Excessive soil in oven.	Instruct user to remove excess before cleaning cycle.

This action is called pyrolysis — chemical decomposition by heat. Any residual ash will be loose and can be wiped out.

The oven is not an incinerator, so do not expect it to burn large quantities of soil. As much excess as possible should be wiped up before the self-cleaning process is started.

In the self-cleaning cycle the oven door must be tightly sealed to limit the entrance of oxygen and it must be securely locked so that people will not try to use it. At these temperatures, burning would be severe and an excess of oxygen into the oven could cause an explosion.

Exact operating sequence for the self-cleaning cycle will vary from model to model. Therefore you must familiarize yourself with the manufacturer's procedure used in the various ovens.

A typical basic sequence for starting the cleaning cycle is as follows:

1. Clean out excess soil, and, if necessary, remove racks (older ovens).
2. Where required, close the door window shutter.
3. Close door and move lock handle, usually to the right.
4. Set the timer, for 2 hours of operation, longer for heavier soils. If there is no timer, the operation must be manually terminated after 2 hours.
5. Set the oven temperature control to clean.

When the oven heats to about 550°F, a lock thermostat will positively lock the door so that it cannot be opened until after the cleaning cycle has been completed and the oven has cooled down to 550°F. This could be an hour after the end of the cleaning cycle. When the lock light has turned off, the oven door can be opened by moving the lock lever back to its original position.

4-8. Electronic Oven

Browning It should be noted here that although the food is completely cooked and ready to serve, some foods need further treatment in order to make them more ap-

Electronic ovens are convenient for cooking food quickly.

pealing to the diner — specifically, meats and seafood, which ordinarily have a seared, browned, or broiled surface finish when cooked in regular ovens.

Browning is accomplished by either pre-broiling the meat before completing the cooking in the microwave oven or by cooking in the oven, then browning to the desired appearance.

Some models include a browning element with its own controls in the microwave oven.

Whether microwaves are reflected or absorbed depends on several factors:
1. The dielectric coefficient of the material
2. Its shape
3. Its mass
4. Its moisture content

Generally metals reflect the wave energy.

Troubleshooting Chart   ELECTRONIC OVENS

PROBLEM	POSSIBLE CAUSE	CORRECTIVE ACTION
No power to oven.	No line voltage. Defective ON/OFF switch. Defective interlock or latch switch.	Check circuit fuses Replace switch Replace switch
Magnetron filament is out.	Open fuse or thermal cut out in filament transformer primary circuit. Open transformer winding. Open Magnetron filament. Loose connection.	Replace fuse of if cut out; check blower for cooling. Replace transformer. Replace Magnetron. Check all wiring.
Unit will not advance to cook phase.	Defective time delay. dc voltage not available. Cook relay not operating. Door switch does not close.	Replace. Check all parts in the high-voltage dc circuit and replace defective one. Check relay coil and contacts. Check switch action with open and shut door. Adjust or replace.
Cook timer fails to operate.	Defective timer.	Replace timer.
Oven does not heat.	Interlock switch is open. Warm-up time delay relay not functioning. Defective Magnetron. High-voltage rectifiers are defective.	Readjust or replace switch. Replace delay. Replace Magnetron. Replace rectifiers.
Oven turns off too soon. Blower continues to operate.	Open thermal protector. High-voltage transformer defective. Defective cook timer.	Check Magnetron cooling system. Replace transformer Replace cook timer.
Power output is low during cook cycle.	Low-line voltage. Shorted diode rectifier in high voltage. Defective Magnetron, if plate current is low and power supply is OK.	Check wiring from panel board to oven. Replace rectifier. Replace Magnetron.

Glass, paper, and most plastics allow the waves to pass through them without absorbing the energy.

For this reason, the oven and glass or paper plates remain cool and only the food gets cooked. Since the food cooks all the way through at once, it can be placed in the oven and can be ready to serve in seconds — but don't try it with TV dinners in the aluminum foil wrapper and plate. *Never use metal containers for the food in the oven.* They will reflect the radio waves away from the food. *Always use glass, plastic, or paper.*

Never operate the oven without some load — the energy has to be absorbed.

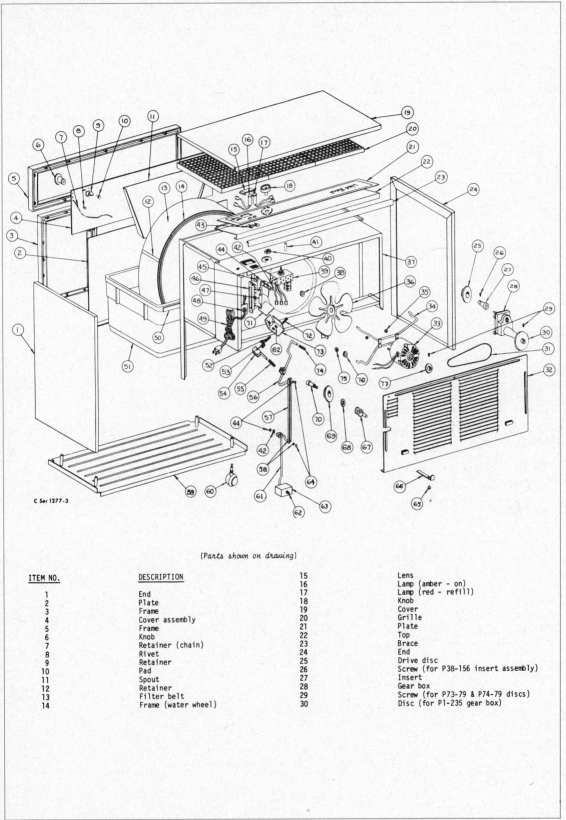

C Ser 1277-3

(Parts shown on drawing)

ITEM NO.	DESCRIPTION		
		15	Lens
		16	Lamp (amber - on)
		17	Lamp (red - refill)
1	End	18	Knob
2	Plate	19	Cover
3	Frame	20	Grille
4	Cover assembly	21	Plate
5	Frame	22	Top
6	Knob	23	Brace
7	Retainer (chain)	24	End
8	Rivet	25	Drive disc
9	Retainer	26	Screw (for P38-156 insert assembly)
10	Pad	27	Insert
11	Spout	28	Gear box
12	Retainer	29	Screw (for P73-79 & P74-79 discs)
13	Filter belt	30	Disc (for P1-235 gear box)
14	Frame (water wheel)		

Exploded view of humidifier and parts list.

ITEM NO.	DESCRIPTION	ITEM NO.	DESCRIPTION
		60	Caster
		61	Arm
31	Belt (for drive motor)	62	Pad (for P2-206 float & P121-42 bracket)
32	Cover	63	Float
33	Fan motor	64	Nut (for P21-137 lever assembly)
34	Frame	65	Rivet
35	Nut (for P24-201 motor assembly)	66	Lever
36	Blade (fan)	67	Nut
37	Chassis	68	Washer
38	Ring (for P12-38 fan assembly)	69	Disc
39	Control - Without exchange	70	Screw
	Control - With exchange	71	Plate
40	Nut (for P2-212 control assembly)	72	Screw
41	Pad	73	Bracket
42	Retainer	74	Pad
43	Nut (for lamp assembly)	75	Nut
44	Washer	76	Grommet
45	Grommet	77	Disc (for P24-201 fan motor)
46	Terminal		
47	Capacitor		
48	Clamp		
49	Cord		
50	Support		
51	Reservoir		
52	Spring		
53	Lug		
54	Switch		
55	Pad		
56	Arm		
57	Lever		
58	Grommet		
59	Bottom		

Parts list of humidifier (continued).

Safety Features Microwaves are no respecters of persons. They would just as soon cook a hand or an arm as a steak. To prevent injury to the user, manufacturers have incorporated a number of safety interlock switches that make it virtually impossible for the user to get hurt or for the oven to get seriously damaged by careless usage.

In addition to user safety, the manufacturers have done much to contain the radio waves *inside* the oven. The glass window has a built-in screen to prevent the radio energy from escaping, and the door gasket must be conductive all around the door to complete the shielding. Whenever any work is done on door or hinge adjustment, always check around the gasket with a strip of paper (a dollar bill is good). There must be a noticeable drag on the paper when you pull it from under the gasket.

Operation Each make of electronic oven will vary from the others in some details of operation, but, in general, they all follow the same sequence.
1. Set food in the oven.
2. Latch door.
3. Turn on the power switch. This step will start the cooling fan and energize the filament of the magnetron.
4. Set the timer for the desired length of cooking time.
5. After a time delay of about a minute to allow the magnetron filament time to warm up, the cooking will commence.
6. Cooking is automatically terminated by the timer. Power is turned off.

4-9. Humidifier

The air-moisture content in a house drops rapidly as the outside temperature drops unless some way is found to put more moisture into the air. Relative humidity as low as 20 percent has been found in houses without humidification or with a humidifier that is not operating correctly.

Low humidity has several effects on a house and its contents. For instance,

- Houseplants will die more readily.
- Rug fibers become brittle and wear faster.
- Boards shrink, thus causing windows and doors to rattle.
- Cracks appear in floorboard seams.
- A piano will go flat.
- Leather shoes dry out.
- Food, if uncovered, dries out.

● Wooden furniture becomes unglued.
● Your nose feels dry and you get colds more often.
● You create static by walking across a carpet.
● You feel the need to turn the thermostat as much as 5 percent higher than normal and still feel cold.

These are all signs that indicate the need of a humidifier. Proper humidity will save money on the heating bill and will provide a healthier environment.

The primary purpose of a humidifier is to put moisture into the air. This might easily mean a gallon of water per day per room. On a very cold day it could mean even more, possibly as much as a gallon per hour for the average house.

Many humidifiers cannot handle that big a job, so do not condemn the machine if it does not produce during extremes of temperature. An owner should be grateful that it is putting *some* moisture into the air. Almost the only certain way of knowing how well a humidifier is doing is by using a sling psycrometer or a wet and dry bulb thermometer with a chart to convert the readings to percent of relative humidity.

Lacking the technical aids just mentioned, you can still tell if the air is too dry if you get a static shock when you walk across a carpet and touch metal. On the other hand, if you fix an iced drink and the glass sweats, the air is humid enough.

Moisture can be put into the air in a number of different ways. These methods range from setting a pan of water in a room and allowing it to evaporate to spraying a fine mist of water into a fan's air stream.

Hard water — in fact, water containing any minerals — is an enemy of the humidifier. If you know that the water is hard (even though a water softener is in use), it is good practice to clean out the water reservoirs of all appliances. Do the same for the associated water-dispensing parts. Developing the habit of doing so will save much future servicing.

If someone fails to fill the reservoir, the unit will cycle on the limit switch, but there will be no humidification of the air.

Remind the user to check the water level regularly! A humidifier can easily use 5 gallons of water in a 24-hour day.

In operation, a water pump raises water up to the media wheel, filter belt, or evaporator pad to be evaporated. A fan then disperses the moist air into the room.

With a bad water situation, it is possible for the water impeller or the tube to become clogged. The result will require dismantling and cleaning.

The media in another type uses a moving belt having one end immersed in a tank of water and the other end exposed to air being circulated by a fan.

Do not be too hasty in condemning the humidistat if the unit runs too long or cycles on too frequently. Make certain that the unit has enough water, and then assume that it is doing its best to put the correct amount of moisture into the air. About the only time to suspect the humidistat is if the humidity is persistently low with the control set to maximum and then the unit is ON for only short intervals. Even then, make certain that water is pumped to the media and is being evaporated into the room. In other words, be sure that the limit switch has not assumed cycling control and that the water supply has been amply maintained for a considerable period of time.

4-10. Dehumidifier

At certain times during the year, some geographic areas are too humid for comfort. If you live in one of these regions, you will probably want a dehumidifier. There may be times when the dehumidifier cannot seem to do the job, when it just is not removing the moisture fast enough.

Be slow to blame the appliance. Check its working conditions first. These include size of room, relative humidity, ventilation. It is possible that weather or household conditions are producing more moisture than is normal. Good spot ventilation, such as exhaust fans, that will remove excessive moisture-ladened air, as it is produced in the laundry, kitchen, and bathroom, will do a great deal to decrease the working load on a dehumidifier.

Basically there are two types of dehumidi-

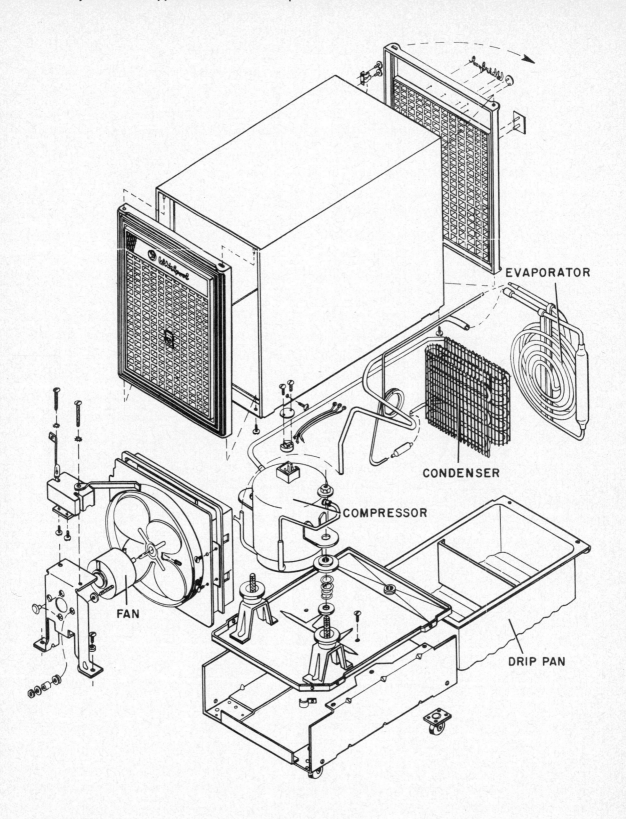

EVAPORATOR

CONDENSER

COMPRESSOR

FAN

DRIP PAN

The air flow is straight through the cabinet in a typical dehumidifier. It passes over the cooling coil (evaporator) to give up its moisture. Then it passes over the condenser and compressor to keep them cool.

fiers. One operates on the absorption principle, drawing air through a chemical drying agent that soaks up the moisture. When it has picked up all the moisture it can hold, this chemical is dried out (regenerated) by an electric heater. This type of machine is not portable in that it requires an outdoor vent. For an equal capacity, it will cost more than the refrigerant type. Its prime advantage is that it will effectively dry air to as low as 20 to 30 percent of relative humidity — even at low temperatures!

The other type of dehumidifier is the refrigerant type. This unit operates on the principle that moisture will condense on a cool surface. Here the fan draws the air over a cool coil of tubing, and the condensed moisture either collects in a pan, which must be emptied regularly, or it runs through a drain. This machine is very efficient on hot, humid days, but its efficiency drops noticeably as the surrounding temperature decreases. For example, at 90°F a machine may remove 3 gallons of water per day, at 80°F it will remove about 2 gallons, and at 70°F it will remove only about 2 quarts of water. Also, this machine is not able to reduce the relative humidity much below 40 to 50 percent because the coils would have to be so cold that the moisture would freeze on them.

The amount of air a dehumidifier can handle depends on its size and local conditions. Most units are rated between 10,000 to 15,000 cubic feet of air. Thus a 12,000-cubic foot unit could handle an 8-foot-high area with 1500 square feet of floor space, or an average ranch-type house. If you find that a dehumidifier is large enough for a home but does not seem to be doing the job, you must consider the sources of humidity.

Trying to dry the air often becomes a major project, like bailing out a leaky boat! Walls, windows, doors, or any source of fresh air coming into the home will allow moisture to enter with it. In addition, there is the steam from cooking, shower, clothes drying, and so on. Consider all these factors before checking the dehumidifier itself for trouble.

To check for humidity, you need a wet and a dry bulb thermometer and a relative humidity chart. Both come in a special tools

category. Whether humidity is too high can also be determined in other ways.
- Doors and drawers swell and stick.
- Shoes and other leather items mildew in the closet.
- Cold-water pipes sweat (drip water).
- Tools get rusty.
- Exterior paint blisters on siding.
- The piano goes sharp in tone.
- Foods mold quickly.
- Last, but hardly least — YOU feel hot and sticky!

Check the owner's manual for specific information on the temperature-humidity range and capacity of a dehumidifier. Also, be sure to follow the manufacturer's advice on emptying the water pan.

Incidentally, this water is completely mineral free and can be used in a steam iron or in the car's storage battery. Make certain that it has not collected dust from the air. If it has, filter it out.

A dirty or clogged condenser will cause poor performance. It should be inspected and vacuum cleaned, or brushed periodically to remove any accumulation of dust.

A dehumidifier should be so located in the room that its end grilles are free of obstructions (at least 6 inches) to allow for unhampered air circulation. The grilles should be cleaned when the coil is cleaned.

Only air that passes through the dehumidifier will be dried. That is why its location in the room is important.

Testing

A dehumidifier is quiet under normal operation, making no more noise than a fan. Any rattles or unusual noises due to vibration can be stopped. Loose wires can be taped to the nearest firm part away from the fan. Tubing can be gently bent away from the parts it might hit. The fan blade must be balanced.

Automatic control is attained with a humidistat, which consists of a snap-action switch actuated by human hair or by a plastic ribbon. Moisture affects the element by causing it to shorten with decreasing humidity and to stretch with increased humidity.

The relative humidity that is to be maintained is set by rotating a knob, thus controlling the linkage between the element and switch. The humidistat is reasonably trouble-free, so don't be too quick to blame it. Check other operating conditions first.

To check the humidistat, disconnect its leads from the humidifier and connect them to a test light or ohmmeter. Rotate the knob from the WET, or 80 percent marking, toward the DRY, 20 percent marking, until a click is heard in the switch. From this point to the DRY end there should be continuity. Rotate the knob back to the WET end until a click is heard in the switch. From this point to the WET end there should be no continuity. Repeat this process several times to be sure that the switch is operating correctly. Note the relative humidity reading at the time the click is heard. This reading should be within + or −10% of the actual room humidity.

If the preceding test shows a defect, replace the humidistat.

Troubleshooting Chart   DEHUMIDIFIERS

PROBLEM	POSSIBLE CAUSE	CORRECTIVE ACTION
Unit does not run.	No power to the unit.	Check fuse, wiring, and outlet.
	Defective motor-starting relay.	Replace relay.
	Humidistat set wrong.	Adjust for desired humidity.
	Faulty humidistat.	Replace humidistat.
	Defective motor or compressor.	Replace refrigerant unit.
Unit turns off and on too frequently.	Defective humidistat.	Replace humidistat.
	Failure in refrigerant system.	Check compressor.
Unit runs but does not dehumidify.	Abnormal conditions.	Check operating conditions.
	Poor location and air circulation.	Relocate with more clearance.
	Defective fan motor.	Replace fan motor.
	Refrigerant low in system.	Locate leak, repair and add more refrigerant.
Unit runs but evaporator frosts.	Abnormal conditions.	Check operating conditions.
	Poor location and air circulation.	Move unit.
Insufficient cooling, both fan and compressor running.	Excessive load due to doors and windows being open; above normal temperature, etc.	Check room conditions.
	Partially clogged filters or air passages.	Clean or replace filters.
		Clean air passages.
	Check duct air damper position	Check control's position and owner's manual for correct setting.
	Fan motor speed set too low.	Check fan speed and adjust if a variable speed fan is used.
	Fan dirty, blades loaded with dust.	Clean fan and blades; make certain fan blades are tight on shaft.
	Compressor not pumping at full capacity.	Replace compressor.
	Insulating seals out of place.	Replace insulation.

Troubleshooting Chart   DEHUMIDIFIERS (continued)

PROBLEM	POSSIBLE CAUSE	CORRECTIVE ACTION
Fan not running.	Defective fan motor. Jammed fan blade. Motor relay defective.	Replace fan motor. Straighten fan blade. Replace motor relay.
Noisy operation.	Fan blade hitting. Tubing hitting. Loose cabinet, etc.	Straighten fan blade. Rebend tubing to clear object being hit. Tighten loose parts that might vibrate.
Unit does not run.	No power to unit. Low voltage. Loose wiring. Defective switch. Defective motor overload protector. Defective fan motor. Defective compressor.	Check fuse, outlet, cord, and cord plug. Overloaded branch circuit. Incorrect wiring. Incorrect voltage for machine's rating. Check all wiring connections. Replace switch. Replace protector. Replace fan motor. Replace compressor.
Fan runs but compressor does not.	Defective compressor. Defective overload protector. Motor capacitor defective. Thermostat set too warm. Thermostat defective. Defrost thermostat defective. Defective switch.	Replace compressor. Replace overload protector. Replace capacitor. Reset thermostat. Replace thermostat. Replace defrost thermostat. Replace switch.
Compressor runs but fan does not.	Defective switch. Fan-speed control reactor defective. Defective fan motor. Defective fan motor capacitor. Fan blades or shaft binding.	Replace switch. Replace reactor. Replace motor. Replace capacitor. Straighten blades or free shaft.
No cooling; both fan and compressor running.	Clogged air passages. Compressor not pumping (coils will not be cool).	Clean filters. If coils are iced, allow to defrost. Clear any foreign obstructions from air passages. Check sealed system.

Preventive Maintenance

The average home has about twenty to thirty electrical appliances. If each appliance had only one failure per year, you would expect then to have an average of a failure every other week. Clearly this is too many, but fortunately most appliances, even when abused, give good performance for much longer than a year. Nevertheless, when an appliance does fail, it is always inconvenient, and it is all the more painful because the failure probably could have been averted or at least delayed for many months by using proper operating procedures and suitable *preventive maintenance.*

If you have to call a repairman, you can expect to pay a bill which is large compared to the original cost of the appliance, even to replace a part costing only a few cents. In view of the high cost of repairs, it is truly amazing that more than one in ten service calls is to replace a burnt-out fuse, to tighten a loose fuse or to plug in an appliance because the user didn't realize the plug had been pulled out. If you are able to do your own appliance repair work, you will solve the fuse problem or plug in the appliance in situations like these, and you can then congratulate yourself on how much money you've saved. It takes little effort, however, to make sure that the fuses (or interlocks) are good in the first place, and that line cords of frequently used appliances remain plugged in. By the same token, the amount of extra effort needed to operate appliances correctly and to keep

them in operating condition is trifling compared to the extended period of trouble-free operation which results.

One of the greatest causes of failure in an electric appliance is not a fault of the appliance at all, but a failure in the line cord or plug. After reading Chapter 3, Volume 4, you know what types of failure you might expect in line cords and plugs, and hopefully you now know how to fix them. However, almost all of these failures can be prevented. Here are some reasons why line cords fail:

1. The user yanks the plug out of the outlet by pulling on the cord.

2. The cord is bent sharply around an obstruction, or a piece of furniture is pushed against a plug, causing a sharp, right-angled bend in the cord.

3. The appliance is dangled by the cord.

4. The cord is run behind a radiator or along a window sill in bright sunlight.

5. The cord is run under a rug and is stepped on frequently.

6. Someone steps on a plug, cracking it.

This is by no means a complete list, but it should give you a good idea as to the causes of cord failure. Also, it is not difficult to figure out how most of these troubles can be prevented. The deceptive part of this problem is that even when the cord is abused frequently, the appliance continues to work well, so it seems hard to justify the extra effort needed, for example, to reach over and pull out the plug by grasping the plug itself instead of yanking the cord. Nevertheless, every time the cord is yanked it shortens the time to the next failure.

When you buy a new appliance, read the instructions before using it. As new improvements are designed into appliances, new methods of operation are needed. It is important for the user to know the limitations of an appliance as well as its advantages. Unfortunately, salesmen stress the advantages and gloss over the limitations. When you buy an appliance, it is best to follow the manufacturer's instructions to the letter. However, auxiliary supplies are also improved, and some changes may be in order. For example, if your dishwasher requires one teaspoon of detergent (according to the instructions supplied with the machine), it may be desirable to use only 1/2 teaspoon of a new, improved detergent.

An electric blanket should give many years of service when used according to the manufacturer's directions. The most common source of trouble in an electric blanket is overheating caused by leaving a weight on the blanket while it is on. The weight may be only a book which you intend to read in bed, but if it rests on the blanket too long, the heat under it cannot escape and may burn the insulation or the wire itself. Also, if an electric blanket is "tucked in", it is bunched or folded so that heat cannot escape. It should be obvious that you should never sit on an electric blanket, even when it is off. The heavy weight on the fragile wires may cause them to break. In summary, electric blankets will give long life, if you remember these restrictions:

1. Don't sit on the blanket.

2. Don't put a weight of any kind on the blanket when it is on. This includes a bedspread over the blanket.

3. Don't tuck in the corners.

4. When the blanket is folded and put away, don't place any weight on top of it.

Where applicable, these same restrictions apply also to heating pads.

If a little is good, a lot is *not necessarily better. This is an important rule to keep in mind when using detergents in dishwashers and clothes washers. Too much detergent can harm an appliance, although the new low-suds detergents cause less trouble than the old sudsing varieties. If you have soft water in your pipes, you should probably use only half the amount of detergent that is recommended by the manufacturers. It is well to experiment when you first get your washer, and use as little detergent as possible to do a satisfactory job. Special rules for dishwashers are:*

1. Make sure the water is at or above the minimum temperature specified by the manufacturer.

2. Don't use the dishwasher as a garbage disposer. Dishes should be scraped first.

3. Learn how to clean the filter and keep it clean.

4. Make sure silverware cannot interfere with the operation of the dishwasher.

Garbage disposers are real work savers, but are abused frequently. They cannot dispose of everything. It goes without saying that spoons, forks, and other items of silverware should not be dropped into the disposer. Items which are not satisfactorily ground include fibrous materials such as corn husks, watermelon rinds and celery. Avoid raw bones. Although these would be ground up eventually, they put too heavy a load on the machine. If the disposer jams, don't panic. Use a long wooden stick to turn the turntable backwards and loosen whatever caused it to jam. You may have to press the restart button to start it again.

Toasters require some special attention. As bread is toasted, pieces frequently break off and fall inside the case. In most cases, small pieces and crumbs fall all the way through

A simple way of improving the efficiency of a heater is to shine the reflector.

and may be removed through an access door. Larger pieces, however, can get caught in the body of the toaster and jam the mechanism or catch fire. It is important then to make sure the toaster is kept clean. The access door should be opened and all crumbs removed periodically. Once a month is not too often. Larger pieces should be removed whenever you know they are there. *Make sure the plug is disconnected whenever you clean out the toaster. Also, make sure you plug it in again when you finish cleaning it.*

A toaster probably takes more abuse than any other appliance. Never slam the handle down. This may bend vital parts, and although the toaster still works, its useful life is shortened.

Refrigerators and freezers should give many years of service. Some important steps to help achieve this are:

1. Make sure the condenser coils are kept clean by vacuuming them at least every six months.

2. Do not keep the refrigerator or freezer in a warm place, as it will have to work harder.

3. Don't pile things around or on top of the unit, as it works better when air circulates freely around it.

4. When the weather is warm or when an extra amount of food is in the cabinet, turn up the coldness control.

5. Check the door gaskets periodically and change them before they become brittle with age.

Motors in many appliances are sealed, self-lubricating units and need no maintenance. In others they should be oiled once or twice a year. As a rule of thumb, if there is an oil cup on a motor or blower or other moving part, you should oil it. But don't flood it with oil. Two or three drops each time are sufficient.

Dirt causes problems in all electric appliances. It was already mentioned that the condenser on a refrigerator should be vacuumed periodically. Similarly, clothes dryers, air conditioners and other large appliances should be checked for dirt and lint. It is a good idea to remove the sides of the cabinet and vacuum out any dirt you can see. Also, filters should be cleaned or changed according to the manufacturer's instructions. For air conditioners and gas furnaces, filters are cheap and should be changed about once a month.

Vacuum cleaners are used to suck up dirt and dust, and consequently, it is surprising that they, too, must be kept clean. Usually, some lint or dirt remains in the hose when the vacuum cleaner is put away. After a few months the lint in the hose limits the amount of air that can enter. One way to prevent lint from remaining in the hose is to run the vacuum cleaner, sucking in clean air, for a few minutes after finishing each day's cleaning. This will cause the lint in the hose to be sucked into the bag. In some cleaners, it is possible to attach the hose to the blower end of the machine, and actually blow out the lint.

With some attention to preventive maintenance, appliances will last for many years. In fact, you may reach a point where you hope an appliance fails so that you can buy a new, improved model. Don't despair because your appliances last too long. If the new model is that attractive, you can donate the old one to a PTA rummage sale, and feel virtuous when you buy the new one.

6 Safety

6-1. Electricity Can Be Lethal

The fact that electricity can kill you is something you must keep in mind whenever you work on any kind of electrical appliance. You might be lucky and escape injury if you carelessly touch a live wire. You might also avoid an accident if you drive an automobile the wrong way on a one-way street. Common sense dictates that both actions should be avoided. Many of the warnings in this chapter also appear in the first chapter, but because of the danger involved, they bear repeating.

Always approach the repair of an electrical appliance with caution. This means giving the task your undivided attention and thinking constantly of the danger. A shock caused by carelessness is just as dangerous as one due to ignorance. You must also bear in mind that after the appliance is repaired, it must be safe to operate. That is, there must be no danger of shock to the person who uses the appliance in a normal manner. There is nothing you can do about someone who uses a metal knife to pry out a piece of bread stuck in a toaster while the toaster is on, but you *can* make sure that he won't get a shock if he touches the case of the toaster and the water faucet at the same time.

There are three simple rules of safety when you are working on appliances. If they are followed *at all times*, you will avoid danger to yourself and to the user.

1. *Do not touch any bare wire* which is connected to an electrical outlet.

2. *Never work on anything "live".*

3. *Always* check to make sure that there is no voltage between the case of the appliance and ground.

The first rule is fairly obvious and is usually carefully observed by beginners, but it is sometimes disregarded by sophisticated repairmen who should know better. It should never be disregarded. *Never* touch a bare wire which is connected to a source of voltage even if you *know* that it is safe. Consider the simulated appliance shown in Figure 2-1.

The line cord is plugged into a receptacle, but the switch in the appliance is in the *open* position. Point A to the left of the switch is now connected directly to the hot side of the line. Thus, point A has a potential of 115 volts from ground. Point A is a dangerous point to touch, especially if you may also be touching ground. (If you are standing on a concrete floor in leather-soled shoes, you are touching ground.) However, you know that point C is connected to the ground side of the line and should be safe to touch. Also, point B is

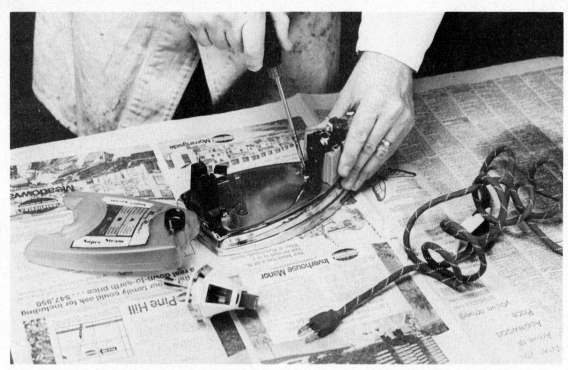

When working on an electrical appliance, always make sure the plug is in plain view.

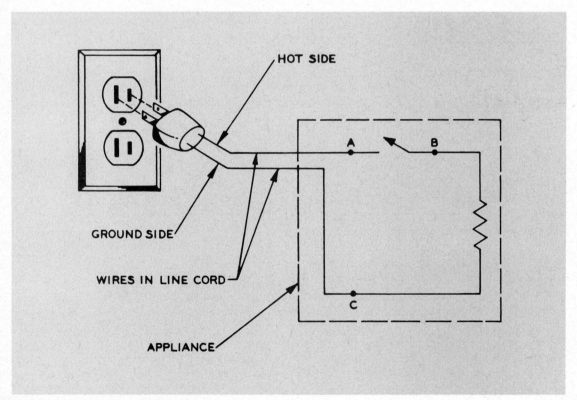

Fig. 6-1. Simulated appliance.

connected to point C through the appliance and therefore should also be safe to touch. The trouble is that you know too much. It is true that you can safely touch points B and C in Figure 2-1. This, however, is a bad habit to get into. In the first place, you might make a mistake as to which is the hot side of the line. Secondly, you might pull out the plug and then put it back upside down so that point A is now grounded, and points B and C are hot. Worse, someone else might do this while your attention is elsewhere. Therefore, never touch a bare wire unless it is disconnected from all voltage sources.

When you are making voltage checks with a meter or a test-light, it is necessary to hold the probes in contact with points of high voltage in the wall outlet or in the appliance being tested. Always hold the probes by the insulation. Again, never touch a bare wire or a bare probe.

The second rule, never work on a "live" appliance, refers to actual repairs. Of course, you must plug the appliance in and turn it on in order to make voltage checks on it. In this sense, you are working on a live appliance. However, once you have located the trouble, you must pull out the plug before making a single change or removing a single screw. Note that it is not enough to shut off the switch. If the appliance is plugged in, there are points of high voltage, (for example, point A in Figure 2-1). Make it a habit to pull out the plug and *have it lying in plain view* while you are working on the appliance. In this way you will not get a shock because you *thought* you pulled the plug.

The third test, the ground test, checks the appliance for safety to the user. The purpose of this check is to make sure that the case of the appliance does not have a voltage on it. Before you do this, it is well to check your outlets to determine which side of the line is grounded. If the receptacle has been wired correctly, the longer slot is the grounded side of the line, as shown in Figure 1-4. However, sometimes a home owner decides to install a new receptacle himself and does so incorrectly. Then the short slot becomes the grounded side. Using a test-light or meter, put one probe in a slot in the receptacle and touch

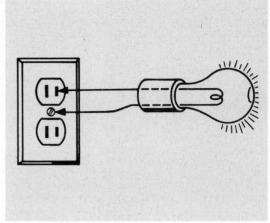

Fig. 2-2. Ground screw.

the other probe to a good ground connection, such as a water pipe. If the bulb lights or the meter reads about 115 volts, the probe in the receptacle is in the hot side of the line. Check the other side, anyway. The light will not go on if you have the wall probe in the slot connected to the grounded side of the line.

If a wall outlet is installed correctly, the mounting screw in the center is connected to ground through the shield that encloses the wires in the wall. Thus, the screw can be used as a ground connection when a water pipe is not available. This is illustrated in Figure 2-2. One probe can be placed on the screw and the other in the hot slot, and the bulb will light. You must be sure that the screw is not covered with paint or other insulating material.

To make the ground safety test, plug in your appliance and check to see if you have voltage between the case and a good ground. If you have ascertained that the screw on the wall outlet is a good ground, you can use that; otherwise, use a water pipe or faucet. The test is illustrated in Figure 2-3. A toaster is shown in the figure, but the same test could be used for any appliance. One probe of the meter is held in contact with the case of the appliance and the other probe is connected to the good ground. In the figure, the screw in the center of the receptacle is used as a ground.

If there is any voltage reading at all, even as little as one volt, there is a potential danger. A low voltage won't give you a shock, but it does

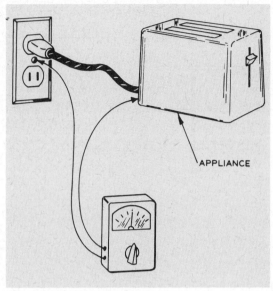

Fig. 2-3. Ground test.

indicate a faulty connection or weak insulation, and it can only get worse. The trouble should be corrected before it becomes dangerous. Note that a normal test-light may not be sensitive enough to detect a low voltage, but a neon tester will. If you do not have a meter, a neon tester is a necessary tool along with a test-light.

In making the ground test, make sure that you pull out the plug and invert it so that the grounded wire in the line cord becomes the hot wire, and vice versa. Make the test for both positions of the plug with the appliance switched on and with it off. There should be no voltage between case and ground in any of these tests. If there is a voltage in even one position of the plug and switch, the appliance is not ready for use until the cause of the trouble is eliminated.

Large appliances such as clothes dryers should be grounded to a water pipe.

One way of making certain that the case of an appliance will never be hot is to permanently fasten the case directly to ground. This should be done on all large appliances that are never moved, such as an oven, clothes washer or clothes dryer. If you have a wall receptacle with a ground terminal as shown in Figure 1-5, then it is a simple matter to run a three-wire line cord to the appliance. The third wire is fastened to the case at the appliance end and to the round prong on the plug. Even without this type of grounding outlet, you can still ground the appliance by running a wire from the case to a ground connection. Use a piece of wire which is rated to carry the current drawn by the appliance. Loosen a screw on the case (preferably in back and out of the way), wrap an end of the wire around the screw, and tighten the screw again. Connect the free end of the wire to a grounding clamp which is attached to a water pipe. Grounding clamps are available in hardware or electrical supply stores. The case of the appliance will then be firmly connected to ground. If subsequently something shorts inside the appliance, putting a voltage on the case, it may cause a fuse to blow, but it will not endanger the life of someone touching the case.

Repairing electrical appliances must not be done haphazardly. Neatness is important, not only because it looks good, but because it is safer. For example, a clean connection has no stray wire sticking out which may contact the case of the appliance or touch another wire causing a short circuit. Connections should be "solid" so that the appliance will stand the abuse that careless users too frequently give them. Although everyone knows that a line cord should not be yanked out of an outlet, people do so frequently. Consequently, wiring must be done in such a way that this type of abuse will put no strain on the electrical connection. It is also obvious that wires must not interfere with moving parts in appliances.

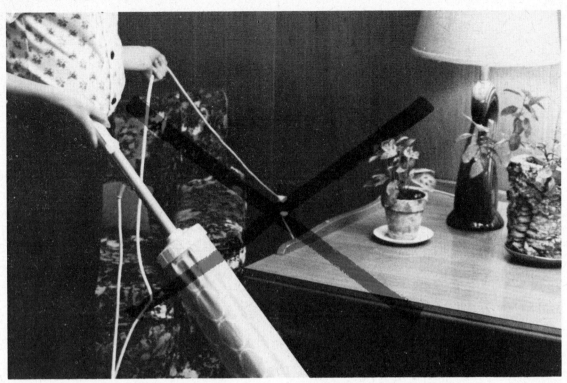

Never yank a cord from the outlet as this will eventually damage the plug and the cord, and create a hazardous condition.

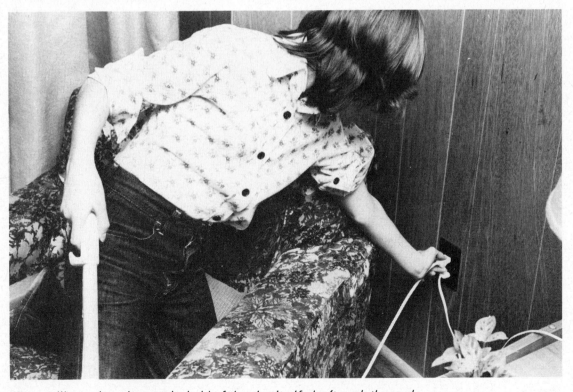

When pulling a plug, always take hold of the plug itself; don't yank the cord.

Some safety hints for the home

The proper way to pull out a plug.

Never pull out a plug by the cord.

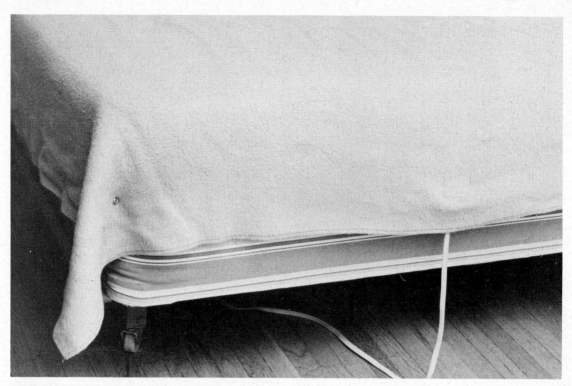

This is the way an electric blanket should be placed on a bed.

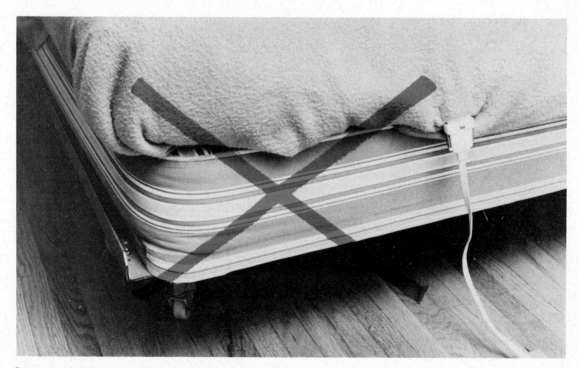

Do not tuck in any part of an electric blanket, as this may cause the elements to break.

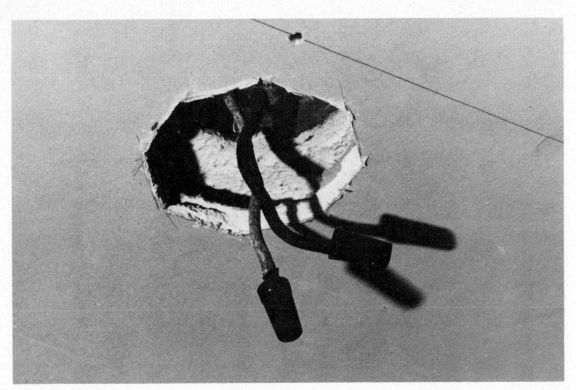

Always cap exposed wires.

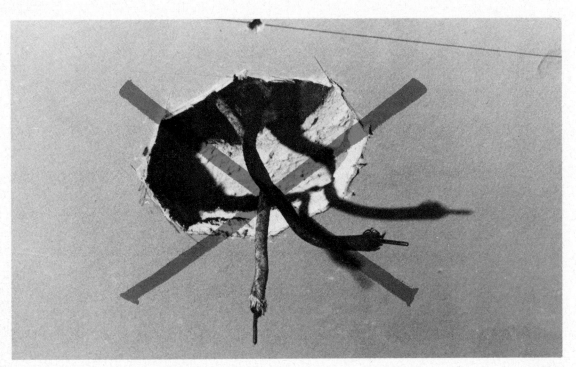

Never leave wires exposed like this.

An "octopus" of extension cords is very dangerous.

An unsafe shortcut.

This cord short-circuited inside the plug and an explosion occurred when the cord was pulled.

A very high voltage current passes through the cord of an electric kettle. Check for wear at the plug and where the cord enters the kettle.

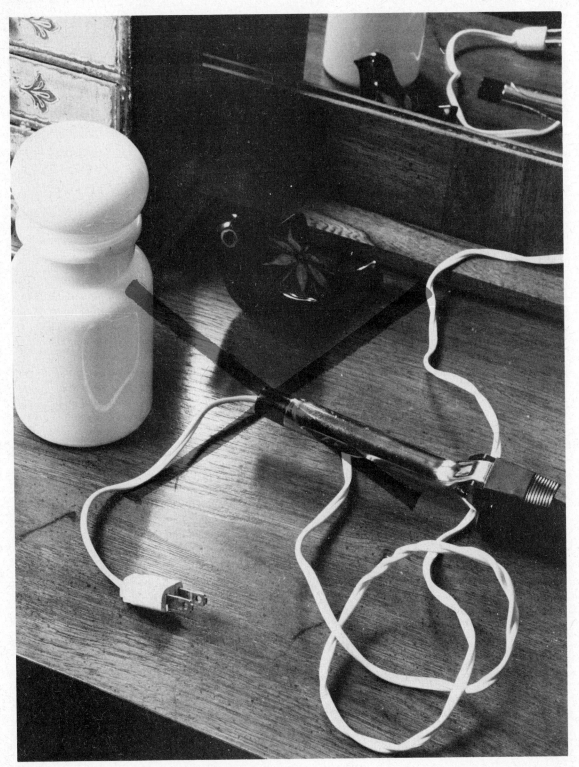

Twisted cords can wear and cause a short circuit.

The small waste receptacle of an electric broom requires emptying frequently.

Clean the lint trap of your vacuum cleaner.

If something appears to be wrong with your stove, the first thing to do is check the fuses.